**Advanced Courses in Mathematics
CRM Barcelona**

Centre de Recerca Matemàtica

Managing Editor:
Carles Casacuberta

Manuel Ritoré
Carlo Sinestrari

Mean Curvature Flow and Isoperimetric Inequalities

Editors for this volume:
Vicente Miquel (Universitat de València)
Joan Porti (Universitat Autònoma de Barcelona)

Birkhäuser
Basel · Boston · Berlin

Authors:

Manuel Ritoré
Departamento de Geometría y Topología
Facultad de Ciencias
Universidad de Granada
18071 Granada
Spain
e-mail: ritore@ugr.es

Carlo Sinestrari
Dipartimento di Matematica
Università di Roma "Tor Vergata"
Via della Ricerca Scientifica
00133 Roma
Italy
e-mail: sinestra@mat.uniroma2.it

2000 Mathematical Subject Classification 53C17, 53A10, 53C20; 52A40

Library of Congress Control Number: 2009936024

Bibliographic information published by Die Deutsche Bibliothek
Die Deutsche Bibliothek lists this publication in the Deutsche Nationalbibliografie;
detailed bibliographic data is available in the Internet at <http://dnb.ddb.de>.

ISBN 978-3-0346-0212-9 Birkhäuser Verlag AG, Basel – Boston – Berlin

© 2010 Birkhäuser Verlag, P.O. Box 133, CH-4010 Basel, Switzerland
Part of Springer Science+Business Media
Printed on acid-free paper produced from chlorine-free pulp. TCF ∞
Printed in Germany

ISBN 978-3-0346-0212-9   e-ISBN 978-3-0346-0213-6

9 8 7 6 5 4 3 2 1   www.birkhauser.ch

Foreword

The *Advanced Course on Geometric Flows and Hyperbolic Geometry* took place from March 12th to 19th, 2008. This was one of the activities of the Research Program *Geometric Flows and Equivariant Problems in Symplectic Geometry*, held at the Centre de Recerca Matemàtica (CRM) in Bellaterra (Barcelona) during the first semester of 2008. It consisted of three main courses delivered by professors Zindine Djadli (Université de Grenoble) on Ricci flow, Manuel Ritoré (Universidad de Granada) on isoperimetric inequalities and hyperbolic geometry, and Carlo Sinestrari (Università degli Studi di Roma Tor Vergata) on mean curvature flow. Some complementary activities were organized during the afternoons. These included preparatory lectures by Carlo Mantegazza, two invited talks by John Lott and Peter Topping, and five contributed talks.

This book contains expanded revisions of the lectures on mean curvature flow and isoperimetric problems given by Carlo Sinestrari and Manuel Ritoré respectively. Although it is impossible to give a complete account of these subjects in such a short series of talks, they succeeded in giving the flavor of the subject, the main techniques (for some of the research streams in their respective subjects, giving references for other viewpoints), deep insights into the fundamental ideas and the state of the art with the major current trends on these matters.

Carlo Sinestrari, a PDE specialist, is especially known in the field of mean curvature flow by his joint work with Gerhard Huisken on the nature of singularities and flow with surgery (see the references [46], [47] and [48] in his part of the book). This is a milestone in the work on mean curvature flow following the approach of classical PDE and an explanation of the ideas of this work are part of the contents of his lectures.

Manuel Ritoré is an expert on the theory of surfaces with constant mean curvature and the isoperimetric problem. Just to give one of his contributions to the field, we shall mention the proof of the double bubble conjecture, obtained in collaboration with Hutchings, Morgan, and Ros. Since the main topic of the first trimester of the research program at the CRM was geometric flows, a significant part of his lectures focused on the attempts to apply geometric flows to solve isoperimetric problems, showing their power at some points and their weakness at others.

For more details about the contents of the lectures, we refer to the Preface of the exposition by Ritoré and the Introduction of that by Sinestrari. We shall only warn the reader about the change of notation in the definition of the second fundamental form and the mean curvature. As is natural, Sinestrari follows the convention of analysts and Ritoré that of geometers. As a consequence, these two quantities have opposite signs in the two parts of the book and, moreover, Sinestrari does not divide the mean curvature by the dimension of the hypersurface. The precise notation is explained at the beginning of each course.

We want to express our gratitude to the director of the CRM, Joaquim Bruna, for his support and the facilities he gave us, to the staff of the CRM, indispensable for the development of this course, and to all colleagues who contributed to our organization of the research semester and all the others who contributed with their participation. Concerning this publication, thanks are given to Manuel Castellet, who proposed it and initiated the work for so doing, and to Carles Casacuberta, who finished it.

Of course, special thanks are given to the authors.

We thank the Ingenio Mathematica project of the Consolider program of the Spanish government for providing generous financial support for the organization of this Advanced Course, as well as the Spanish Ministry of Science and Innovation through FEDER/MEC grant MTM2007-29362-E and the Catalan government through AGAUR grant 2007PIV10001.

Vicente Miquel and Joan Porti

Contents

Part I

Formation of Singularities in the Mean Curvature Flow

Carlo Sinestrari

1 Introduction

Let $F_0 \colon \mathcal{M} \to \mathbb{R}^{n+1}$ be a smooth immersion of an n-dimensional hypersurface in Euclidean space, $n \geq 1$. The evolution of $\mathcal{M}_0 = F_0(\mathcal{M})$ by mean curvature flow is a one-parameter family of immersions $F \colon \mathcal{M} \times [0, T) \to \mathbb{R}^{n+1}$ satisfying

$$\frac{\partial F}{\partial t}(p, t) = -H(p, t)\, \nu(p, t), \qquad p \in \mathcal{M},\, t \geq 0, \tag{1.1}$$

$$F(\,\cdot\,, 0) = F_0, \tag{1.2}$$

where $H(p, t)$ and $\nu(p, t)$ are respectively the mean curvature and the normal at the point $F(p, t)$ of the surface $\mathcal{M}_t = F(\,\cdot\,, t)(\mathcal{M})$.

It can be checked that $-H(p, t)\, \nu(p, t) = \Delta_{\mathcal{M}_t} F(p, t)$, where $\Delta_{\mathcal{M}_t}$ is the Laplace–Beltrami operator on $\mathcal{M}_t$. Thus, the mean curvature flow may be regarded as a kind of heat equation for the immersion. In particular, it can be shown that it is a parabolic problem and is uniquely solvable for small times. In addition, the solutions satisfy comparison principles and derivative estimates similar to the case of parabolic equations in euclidean space. This motivates the minus sign in (1.1); otherwise the problem would be backward parabolic and ill-posed in general. However the mean curvature flow is not really equivalent to a heat equation, since the operator $\Delta_{\mathcal{M}_t}$ is not the laplacian with respect to a fixed metric, but it depends on $\mathcal{M}_t$ which is the unknown of the problem. In particular, in contrast to the classical heat equation, the mean curvature flow is a nonlinear evolution and the solutions exist in general only in a finite time interval.

Mean curvature flow occurs in the description of the interface evolution in certain physical models (see, e.g., [12, 54, 62]). This is related to the property that such a flow is the gradient flow of the area functional (see Section 4) and therefore appears naturally in problems where a surface energy is minimized. Another interesting feature of the flow is the connection with certain reaction-diffusion equations. Consider for example the equation

$$\frac{\partial u}{\partial t} = \Delta u - \frac{1}{\varepsilon} W'(u), \tag{1.3}$$

where $W(u) = (u^2 - 1)^2$ (double-well potential). One can study the singular limit of the solutions when ε tends to zero. Under suitable hypotheses, it can be shown that the solution u_ε of (1.3) with fixed initial data converges as $\varepsilon \to 0^+$ to a function which assumes only the values ± 1 in regions separated by boundaries which evolve by mean curvature flow (see [4, 62]).

A further motivation for the study of the mean curvature flow comes from geometric applications, in analogy with the Ricci flow of metrics on abstract Riemannian manifolds. One can use the flow as a tool to obtain classification results for surfaces satisfying certain curvature conditions, or to derive isoperimetric inequalities, or to construct minimal surfaces. As in Hamilton's program for the Ricci flow, a fundamental step to the study of these problems is the definition of a flow with surgeries which allows us to continue the flow through the singularities in a controlled way. We will see that there are several analogies between Ricci flow and mean curvature flow, although there is no direct link between the two.

Many interesting generalizations of the mean curvature flow can be studied. For instance, we can consider other geometric flows of the form

$$\frac{\partial F}{\partial t}(p,t) = -\mathcal{S}(p,t)\,\nu(p,t), \qquad p \in \mathcal{M},\, t \geq 0, \tag{1.4}$$

where $\mathcal{S}(p,t)$ is a given symmetric function of the principal curvatures of $\mathcal{M}_t$ at p, t. A typical example is the Gauss curvature flow, where $\mathcal{S}$ is the product of the curvatures, which has been proposed as a model for the erosion of stones by sea waves in [25]. Further generalizations have been considered, such as speeds containing non-local terms, or depending on curvatures which are defined using anisotropic geometries.

In these notes we first collect the basic properties of the general geometric flows of the form (1.4). Then, we focus on the mean curvature flow and present the results by Huisken and Huisken–Sinestrari [38, 42, 46, 47, 48] about the formation of singularities, curvature estimates and the construction of a flow with surgeries.

An important subject which we do not treat in these notes are the weak solutions. Let us just recall that, in contrast to the Ricci flow, many notions of weak solutions are known for the mean curvature flow. The first one was introduced by Brakke in his pioneering work [12], which employed techniques of geometric measure theory. Another important kind of weak solution, based on the level set approach and the theory of viscosity solutions, was introduced independently by Chen, Giga, Goto [14] and by Evans and Spruck [23]. Further results, including a link between the two approaches, were provided by Ilmanen in [49]. According to these definitions, a solution of the mean curvature flow exists also after the formation of singularities. However, it is no longer a surface but a weaker object; for instance, it is a varifold (a suitable Radon measure) in Brakke's theory, and the level set of a function (which may be singular or have non-empty interior) in the approach of [14, 23]. In addition, uniqueness is no longer ensured in general. Weak solutions have been studied over the years by many authors, who have introduced other notions in addition to the above ones. An important problem in this field is the analysis of singularities of weak solutions, consisting of the estimation of the possible Hausdorff dimension of the singular set. This study employs techniques which are different from the ones presented in these notes for smooth solutions, but the results obtained show sometimes interesting connections, see, e.g., [68]. The reader who is interested in these topics may consult the monograph by Giga [28] to learn about the level set approach and the one by Ecker [21] to find a proof of Brakke's main regularity theorem which uses a simpler procedure than the original one.

We consider in these notes only curvature flows of codimension 1 in euclidean spaces. There are important results about curvature flows where the ambient space is a Riemannian or Lorentzian manifold, see, e.g., [44, 27]. Not many results are known about the case where the codimension is greater than 1; a special case of higher codimension where some interesting results about the singularities have been obtained is the so-called Lagrangian mean curvature flow. The interested reader may consult, e.g., [5, 61, 65, 66] and the references therein.

2 Geometry of hypersurfaces

In this section we recall some basic notions and fix some notation about the geometry of immersed manifolds. A more detailed exposition can be found for instance in [60, 21]. We restrict ourselves to manifolds of codimension 1 in a euclidean ambient space, that is, immersions of the form $F \colon \mathcal{M} \to \mathbb{R}^{n+1}$, where $\mathcal{M}$ has dimension n, for some $n \geq 1$. The manifold $\mathcal{M}$ is assumed smooth and without boundary; it can be either closed, or complete and non-compact. The image $F(\mathcal{M})$ is called a hypersurface; for simplicity, we will often use equivalently the word "surface", regardless of the dimension. Also, we will often identify the points of $\mathcal{M}$ with their image under the immersion, if there is no risk of confusion.

We follow the notation of [38]; in particular, we use the classical tensor notation with coordinates and upper and lower indices, and with the Einstein convention about summing over repeated indices. Let $x = (x_1, \ldots, x_n)$ be a local coordinate on $\mathcal{M}$. The components of a vector v in the given coordinate system are denoted by v^i, while the ones of a covector w are w_i. Mixed tensors have components with upper and lower indices depending on their type.

The Riemannian metric g on $\mathcal{M}$ is given by

$$g_{ij}(x) = \left\langle \frac{\partial F}{\partial x_i}, \frac{\partial F}{\partial x_j} \right\rangle_e,$$

where $\langle \cdot, \cdot \rangle_e$ denotes the euclidean scalar product in $\mathbb{R}^{n+1}$ (the subscript will be sometimes dropped if there is no risk of confusion). The metric induces a natural isomorphism between tangent and cotangent space. In coordinates, this is expressed in terms of raising/lowering indexes by means of the matrices g_{ij} and g^{ij}, where g^{ij} is the inverse of g_{ij}. The scalar product on the tangent space extends to any tensor bundle, by contracting any pair of lower and upper indices with g^{ij} and g_{ij} respectively. Thus, for instance, the scalar product of two $(1, 2)$-tensors T^i_{jk} and S^i_{jk} is defined by

$$\langle T^i_{jk}, S^i_{jk} \rangle = T^{jk}_i S^i_{jk} = T^l_{pq} S^i_{jk} g_{li} g^{pj} g^{qk}.$$

This also allows us to define the norm of any tensor T as $|T| = \sqrt{\langle T, T \rangle}$. In addition, the volume element $d\mu$ on $\mathcal{M}$ is given in local coordinates by

$$d\mu = \sqrt{\det g_{ij}} \, dx.$$

The covariant derivative of a vector v^i or a covector w_j are defined respectively as

$$\nabla_k v^i = \frac{\partial v^i}{\partial x_k} + \Gamma^i_{jk} v^j, \qquad \nabla_k w_j = \frac{\partial w_j}{\partial x_k} - \Gamma^i_{jk} w_i,$$

where Γ^i_{jk} are the Christoffel symbols of the Levi-Civita connection associated with the metric g. Covariant derivatives of general tensors are defined in a similar way, e.g., we have for a $(1,2)$-tensor T^i_{jl},

$$\nabla_k T^i_{jl} = \frac{\partial T^i_{jl}}{\partial x_k} + \Gamma^i_{mk} T^m_{jl} - \Gamma^m_{jk} T^i_{ml} - \Gamma^m_{kl} T^i_{jm}.$$

If f is a function we simply set $\nabla_k f = \frac{\partial f}{\partial x_k}$. Thus, ∇f coincides with the differential df; it can also be regarded as an element of the tangent space, via the isomorphism induced by the metric, and in this case is called the *gradient* of f. The gradient of f can be identified with a vector in $\mathbb{R}^{n+1}$ via the differential dF; such a vector is called the *tangential gradient* of f, is denoted by $\nabla^M f$ and is given by

$$\nabla^M f = \nabla^i f \frac{\partial F}{\partial x_i} = g^{ij} \frac{\partial f}{\partial x_j} \frac{\partial F}{\partial x_i}. \tag{2.1}$$

The word "tangential" comes from an equivalent definition of $\nabla^M f$ in the case where f is a function defined in the ambient space $\mathbb{R}^{n+1}$. Then it can be checked that $\nabla^M f$ is the projection of the standard euclidean gradient Df on the hyperplane tangent to $\mathcal{M}$, that is

$$\nabla^M f = Df - \langle Df, \nu \rangle_e \nu, \tag{2.2}$$

where ν is the unit normal to $\mathcal{M}$.

The *second fundamental form* associated with $F\colon \mathcal{M} \to \mathbb{R}^{n+1}$ is the tensor $A = (h_{ij})$ defined as

$$h_{ij} = -\left\langle \frac{\partial^2 F}{\partial x_i \partial x_j}, \nu \right\rangle_e = \left\langle \frac{\partial F}{\partial x_i}, \frac{\partial \nu}{\partial x_j} \right\rangle_e.$$

The *Weingarten operator* W is obtained from the second fundamental form by raising one index, i.e., it has components $h^i_j = g^{ik} h_{kj}$. The *principal curvatures* of $\mathcal{M}$ at a point are the eigenvalues of h^i_j or, equivalently, the eigenvalues of h_{ij} with respect to g_{ij}. We denote the principal curvatures by $\lambda_1 \le \cdots \le \lambda_n$. The *mean curvature* is defined as the trace of the second fundamental form, i.e.,

$$H = g^{ij} h_{ij} = h^i_i = \lambda_1 + \cdots + \lambda_n.$$

An important quantity in our analysis is the squared norm of the second fundamental form, which is equal to

$$|A|^2 = h^{ij} h_{ij} = \lambda_1^2 + \cdots + \lambda_n^2.$$

It is easy to see that $|A|^2 \ge H^2/n$, with equality if and only if all curvatures coincide; in fact we have the identity

$$|A|^2 - \frac{H^2}{n} = \frac{1}{n} \sum_{i<j} (\lambda_j - \lambda_i)^2. \tag{2.3}$$

Clearly, A, W, H depend on the choice of orientation; if ν is reversed, they are all multiplied by a minus sign. On the contrary, the vector $-H\nu$, called the *mean curvature vector*, is independent of the orientation; in particular, it is well defined globally even if $\mathcal{M}$ is non-orientable.

The reader should be aware that different sign conventions are sometimes used in the literature when dealing with the above quantities. Also, the mean curvature is often defined as $(\lambda_1 + \cdots + \lambda_n)/n$, i.e., as the mean value of the curvatures, rather than the sum, as in the definition we have chosen. Of course, this makes no substantial difference in the analysis.

We say that the hypersurface $\mathcal{M}$ is *convex* if the principal curvatures are non-negative everywhere. Observe in fact that, with our definitions, if $F(\mathcal{M})$ is the boundary of a convex set, and the normal is directed toward the exterior, then all principal curvatures are non-negative.

The Gauss equations allow us to write the components R_{ijkl} of the Riemann curvature tensor in terms of the second fundamental form. We have in fact

$$R_{ijkl} = h_{ik}h_{jl} - h_{il}h_{jk}.$$

Therefore, the scalar curvature R satisfies

$$R = g^{ik}g^{jl}R_{ijkl} = H^2 - |A|^2 = 2\sum_{i<j}\lambda_i\lambda_j. \qquad (2.4)$$

We also recall the *Codazzi equations*, which say that

$$\nabla_i h_{jk} = \nabla_j h_{ik}, \qquad i, j, k \in \{1, \ldots, n\}.$$

Taking into account the symmetry of h_{ij}, the Codazzi equations imply that the tensor $\nabla A = \nabla_i h_{jk}$ is totally symmetric.

We recall that the *Laplace–Beltrami operator* on functions $f\colon \mathcal{M} \to \mathbb{R}$ is defined as

$$\Delta_{\mathcal{M}} f = g^{ij}\nabla_i\nabla_j f = g^{ij}\left(\frac{\partial f}{\partial x_i \partial x_j} - \Gamma_{ij}^k \frac{\partial f}{\partial x_k}\right).$$

We write simply Δ instead of $\Delta_{\mathcal{M}}$ if there is no risk of ambiguity. Like the standard laplacian in euclidean space, $\Delta_{\mathcal{M}}$ is an elliptic operator. On a closed manifold it satisfies the usual identities

$$\int_{\mathcal{M}} f\,\Delta h\,d\mu = -\int_{\mathcal{M}} \langle \nabla f, \nabla h\rangle\,d\mu = \int_{\mathcal{M}} h\,\Delta f\,d\mu$$

for any pair of functions $f, h\colon \mathcal{M} \to \mathbb{R}$. If f is a function defined in the whole ambient space $\mathbb{R}^{n+1}$, then we have the relation (see [21])

$$\Delta_{\mathcal{M}} f = \Delta_{\mathbb{R}^{n+1}} f - \frac{\partial^2 f}{\partial \nu^2} - H\frac{\partial f}{\partial \nu}, \qquad (2.5)$$

where $\partial f / \partial \nu$ and $\partial^2 f / \partial \nu^2$ are directional derivatives of f in the standard euclidean sense. This identity shows that $\Delta_{\mathcal{M}}$ not only neglects the contribution of the normal direction in the standard laplacian, but also takes into account the curvature of $\mathcal{M}$.

In particular, if we denote by $Y = (y_1, \ldots, y_{n+1})$ the points in $\mathbb{R}^{n+1}$, we find

$$\Delta_{\mathcal{M}} y_\alpha = -H\nu_\alpha, \qquad \alpha = 1, \ldots, n+1, \tag{2.6}$$

where ν_α is the α-component of ν. In addition,

$$\Delta_{\mathcal{M}} |Y|_e^2 = 2n - 2H \langle Y, \nu \rangle_e. \tag{2.7}$$

Given $V : \mathcal{M} \to \mathbb{R}^{n+1}$ it is possible to define $\Delta_{\mathcal{M}} V : \mathcal{M} \to \mathbb{R}^{n+1}$ by applying $\Delta_{\mathcal{M}}$ to each of the $(n+1)$ components of V separately. It is interesting to compute $\Delta_{\mathcal{M}} F$, where F is the function which gives the immersion. Since by definition $F(p) \equiv Y$ for any $p \in \mathcal{M}$, equation (2.6) shows that $\Delta_{\mathcal{M}} F = -H\nu$, that is, $\Delta_{\mathcal{M}} F$ coincides with the mean curvature vector.

Given a vector field $V = (V^i)$ on $\mathcal{M}$, the *tangential divergence* of V is defined as

$$\mathrm{div}^M V = \nabla_i V^i. \tag{2.8}$$

On a closed manifold $\mathcal{M}$, the *divergence theorem* says that, for any vector field V, we have

$$\int_{\mathcal{M}} \mathrm{div}^M V \, d\mu = 0.$$

The tangential divergence can also be defined for vector fields which are not tangent to $\mathcal{M}$, i.e., general fields $V : \mathcal{M} \to \mathbb{R}^{n+1}$. Let us denote by $V_1, \ldots, V_{n+1}$ the components of V and by $e_1, \ldots, e_{n+1}$ the vectors of the canonical basis in $\mathbb{R}^{n+1}$. Then $\mathrm{div}^M V : \mathcal{M} \to \mathbb{R}$ is defined as

$$\mathrm{div}^M V = \sum_{\alpha=1}^{n+1} (\nabla^M V_\alpha) \cdot e_\alpha. \tag{2.9}$$

Here the dot is the scalar product in $\mathbb{R}^{n+1}$ and $\nabla^M V_\alpha$ is the tangential gradient of the function V_α. It can be checked that this definition coincides with (2.8) when V is tangent to $\mathcal{M}$. Also (see [60]), the divergence theorem can be generalized as follows: if $\mathcal{M}$ is closed, then for any $V : \mathcal{M} \to \mathbb{R}^{n+1}$,

$$\int_{\mathcal{M}} \mathrm{div}^M V \, d\mu = \int_{\mathcal{M}} H(V \cdot \nu) \, d\mu. \tag{2.10}$$

Let us derive a more explicit expression of $\mathrm{div}^M V$ when V is defined in the whole ambient space and the tangential gradient can be computed using (2.2). If we set

$\nu_\alpha = \nu \cdot e_\alpha$, we obtain

$$\mathrm{div}^M V = \sum_{\alpha=1}^{n+1} \left(DV_\alpha - \left(\sum_{\beta=1}^{n+1} \frac{\partial V_\alpha}{\partial y_\beta} \nu_\beta \right) \nu \right) \cdot e_\alpha$$

$$= \sum_{\alpha=1}^{n+1} \frac{\partial V_\alpha}{\partial y_\alpha} - \sum_{\alpha,\beta=1}^{n+1} \frac{\partial V_\alpha}{\partial y_\beta} \nu_\alpha \nu_\beta. \tag{2.11}$$

We conclude by giving the explicit expression of the main geometric quantities in the case where $\mathcal{M}$ is the graph of a function $x_{n+1} = u(x_1, \ldots, x_n)$. We denote by $D_i, D_{ij}^2, \ldots$ the partial derivatives in $\mathbb{R}^n$ and we choose the orientation of ν which points downward. Then we find, by a straightforward computation,

$$\nu = \frac{(D_1 u, \ldots, D_n u, -1)}{\sqrt{1 + |Du|^2}}, \tag{2.12}$$

$$g_{ij} = \delta_{ij} + D_i u D_j u, \qquad g^{ij} = \delta_{ij} - \frac{D_i u \, D_j u}{1 + |Du|^2}, \tag{2.13}$$

$$h_{ij} = \frac{D_{ij}^2 u}{\sqrt{1 + |Du|^2}}, \qquad H = \mathrm{div}\left(\frac{Du}{\sqrt{1 + |Du|^2}} \right), \tag{2.14}$$

where div is the standard divergence in $\mathbb{R}^n$.

3 Examples

There are very few examples where the solution to the mean curvature flow can be explicitly computed, which we describe in the following.

Example 3.1. The simplest example is given by the evolution of a sphere. Suppose that $\mathcal{M}_0 = S_R^n(0)$, the sphere of radius R around the origin. Then it is easy to see that the evolution of $\mathcal{M}_0$ is given by spheres $\mathcal{M}_t = S_{r(t)}^n(0)$, where the radius $r(t)$ evolves according to the ordinary differential equation

$$r'(t) = -\frac{n}{r(t)}, \qquad r(0) = R.$$

The solution is given by $r(t) = \sqrt{R^2 - 2nt}$. Observe that $r(t) \to 0$ as $t \to R^2/2n$, that is, the sphere shrinks to a point in finite time and the flow is no longer defined for $t \geq R^2/2n$. The evolution of a cylinder $S^{n-k} \times \mathbb{R}^k$ is analogous: the spherical factor shrinks while the flat one remains unchanged.

Analogous computations can be done for a general curvature flow of the form (1.4). For instance, for the Gauss curvature flow the radius of a shrinking sphere is given by

$$r(t) = \left(R^{n+1} - (n+1)t \right)^{\frac{1}{n+1}}, \qquad t \in \left[0, \frac{R^{n+1}}{n+1} \right).$$

The behaviour is different in the case of expanding flows, like the inverse mean curvature flow, corresponding to $\mathcal{S} = -1/H$. In this case the solution starting from a sphere of radius R exists for all times and is a sphere of radius $r(t) = Re^{t/n}$, for $t \in [0, +\infty)$. □

Example 3.2 (Translating graphs). Let us consider an evolving surface which is represented (either globally or locally) as the graph of a function $x_{n+1} = u(x_1, \ldots, x_n, t)$. Using (2.13)–(2.14) it is easily checked that the surface evolves by mean curvature flow if and only if u satisfies the partial differential equation

$$\frac{\partial u}{\partial t} = \sqrt{1 + |Du|^2} \, \operatorname{div}\left(\frac{Du}{\sqrt{1 + |Du|^2}}\right). \tag{3.1}$$

Let us consider translating solutions of the flow, that is, functions such that $u(x, t) = v(x) + t$ for some v. Such solutions are also called "translating solitons". Then v must satisfy

$$\sqrt{1 + |Dv|^2} \, \operatorname{div}\left(\frac{Dv}{\sqrt{1 + |Dv|^2}}\right) = 1. \tag{3.2}$$

In one dimension the equation becomes

$$v_{xx} = 1 + v_x^2,$$

which can be integrated explicitly. A solution is given by $v(x) = -\ln \cos x$, for $x \in (-\pi/2, \pi/2)$, while all other ones are obtained by translating or adding constants. This solution of the mean curvature flow is usually called the *grim reaper*. It can be regarded as the counterpart of the "cigar soliton" in the Ricci flow.

In higher dimension, the analysis of (3.2) is more difficult. It can be shown that there exists a unique convex rotational symmetric solution. Unlike the grim reaper, it is defined in the whole space and has a quadratic growth at infinity. The properties of this solution have been investigated in [17]. The existence of entire convex translating graphs which are not rotationally symmetric has been studied in [67]. □

4 Local existence and formation of singularities

For a geometric flow of the form (1.4) we have the following result, which ensures local existence and uniqueness of the solution under a suitable assumption on the initial data.

Theorem 4.1. *Let $F_0 : \mathcal{M} \to \mathbb{R}^{n+1}$ be a smooth immersion of a closed n-dimensional manifold $\mathcal{M}$. If F_0 is such that, at any point $p \in \mathcal{M}$, we have*

$$\frac{\partial \mathcal{S}}{\partial \lambda_i}(\lambda_1(p), \ldots, \lambda_n(p)) > 0, \qquad i = 1, \ldots, n, \tag{4.1}$$

then equation (1.4) with initial value (1.2) has a unique smooth solution on some time interval $[0, T)$.

Condition (4.1) ensures that the flow (1.4) is parabolic in a neighbourhood of the immersion F_0. Observe that this condition is always satisfied in the case of the mean curvature flow.

For the mean curvature flow, the above theorem was proved using three different approaches by Gage–Hamilton [26], Ecker–Huisken [22] and Evans–Spruck [24]. For a general speed, this theorem is stated in [45]; a proof can be found in [27] (see also [29] for a similar result). Let us give some informal remarks about the strategy of the proof.

In the proof by Gage and Hamilton, one interprets equation (1.1) as a system of partial differential equations in the unknown F. As in the case of the Ricci flow, the system turns out to be parabolic, but not strictly parabolic because of the invariance under tangential diffeomorphisms. However, one can apply the existence result by Hamilton [33] for systems with an integrability condition, and prove well-posedness of the problem.

As in the Ricci flow case, where Hamilton's original approach can be simplified by De Turck's trick, it is possible to find simpler equivalent formulations for the mean curvature flow as a scalar partial differential equation which is strictly parabolic. This is easy if the surface is the graph of a function in $\mathbb{R}^n$; then equation (1.1) is equivalent to (3.1) and can be studied by standard quasilinear parabolic theory. In our theorem we consider closed surfaces which cannot be written globally as graphs. In this case we can represent the unknown surface $\mathcal{M}_t$ for small time as a graph over the initial value $\mathcal{M}_0$. It turns out that the height of the graph satisfies a parabolic equation, for which existence and uniqueness for small times can be proved. This is the technique used in [22, 27].

The method of [24, 29] is inspired by the level set approach. These authors consider the signed distance function from $\mathcal{M}_t$ and show that it satisfies a parabolic equation in $\mathbb{R}^{n+1}$; they are able to prove that the equation has a smooth solution in a neighbourhood of $\mathcal{M}_0$ for small times.

Roughly speaking, the same techniques can be extended to the more general flow of the form (1.4), because it can be checked that condition (4.1) ensures the parabolicity of the equations which are considered in the above approaches. In general these equations will be fully nonlinear, instead of quasilinear as in the mean curvature flow. Fully nonlinear equations are in some respect difficult to treat; for example, a good regularity theory is only known under the additional assumption that the operator is concave (which would correspond to the concavity of the speed S with respect to the curvatures). However, small time existence can be established without further assumptions; it can be proved by a standard fixed point argument in Hölder spaces, see [50, §8.5.3] for a proof in a euclidean setting.

As we have remarked above, condition (4.1) is satisfied in the case of the mean curvature flow for any initial surface. For other flows, the initial surfaces must belong to suitable classes. For instance, the Gauss curvature flow is strictly parabolic only when all λ_i's are positive, i.e., when $\mathcal{M}_0$ is uniformly convex. It is interesting to consider certain borderline cases which are non-strictly parabolic, such as the Gauss curvature flow of a convex surface with a flat part, see, e.g., [19].

Another consequence of the parabolicity of the equation is the following avoidance principle, which we state here in the case of the mean curvature flow.

Theorem 4.2. *Let $\mathcal{M}_0, \mathcal{N}_0$ be two smooth closed surfaces and let $\mathcal{M}_t, \mathcal{N}_t$ be their evolutions under mean curvature flow. Let $T > 0$ be such that $\mathcal{M}_t, \mathcal{N}_t$ are both defined for $t \in [0, T]$. Suppose that $\mathcal{M}_0, \mathcal{N}_0$ are disjoint. Then $\mathcal{M}_t$ and $\mathcal{N}_t$ are disjoint for all $t \in (0, T]$.*

Proof (Sketch). If the result is not true, then there is a first time $t_0 > 0$ at which the surfaces intersect. At any intersection point, the normals to the surfaces must coincide up to the sign. Then we can write locally the surfaces as graphs in a neighbourhood of the intersection point for t close to t_0. By construction, the graphs solve equation (3.1), are disjoint for $t < t_0$ and touch for $t = t_0$, and this is a contradiction with the strong maximum principle. □

A similar result holds for a general flow of the form (1.4) under a suitable parabolicity condition see, e.g., [58]. However, the statement becomes more complicated, because, in contrast to the mean curvature flow, the other flows in general depend on the choice of the orientation. Therefore, when two surfaces touch at a point where the normals point in opposite directions, the corresponding graphs do not satisfy the same equation. In fact, it is easy to see that Theorem 4.2 does not hold in the same generality for other flows: for example, two disjoint spheres evolving by inverse mean curvature flow have non-empty intersection starting from some positive time.

Many geometric quantities associated with a surface evolving by (1.4) satisfy a parabolic equation. Before stating the result, we need some preliminary definitions and remarks. The speed $\mathcal{S}$ governing our flow is by definition a function of the curvatures λ_i's. Since the curvatures are functions of h^i_j, we can also regard $\mathcal{S}$ as a function of h^i_j. The regularity of $\mathcal{S}$ interpreted in this way is not obvious, because the eigenvalues are not a smooth function of the entries of a matrix (they are in general differentiable only as long as they are all distinct). However, here we do not consider the eigenvalues separately, but a function $\mathcal{S}$ which by assumption is symmetric. In this case it can be proved (see [27, Theorem 2.1.20]) that if $\mathcal{S}$ is smooth with respect to λ_i, it is also smooth with respect to h^i_j. Then we can define an operator $\mathcal{L}_\mathcal{S}$, associated with $\mathcal{S}$, as

$$\mathcal{L}_\mathcal{S} f = \frac{\partial \mathcal{S}}{\partial h^i_j} \nabla^i \nabla_j f \tag{4.2}$$

for any $f \colon \mathcal{M} \to \mathbb{R}$. Observe that, if $\mathcal{S} = H$, then $\mathcal{L}_\mathcal{S}$ is the Laplace–Beltrami operator $\Delta_\mathcal{M}$. In general, it can be checked that the operator $\mathcal{L}_\mathcal{S} f$ is elliptic if and only if the surface satisfies condition (4.1) which ensures the parabolicity of the flow.

We restrict for simplicity to the case where $\mathcal{S}$ is a homogeneous function of the curvatures. Then we have the following equations, which can be derived by computations analogous to those in [38, §3]; see also [6, §3].

Lemma 4.3. *If $\mathcal{M}_t$ evolves by (1.4), with $\mathcal{S}$ a symmetric function of the curvatures homogeneous of degree $\beta > 0$, then the geometric quantities associated to $\mathcal{M}_t$ satisfy the following equations:*

(i) $\dfrac{\partial}{\partial t} g_{ij} = -2\mathcal{S} h_{ij}, \qquad \dfrac{\partial}{\partial t} g^{ij} = 2\mathcal{S} h^{ij},$

(ii) $\dfrac{\partial}{\partial t} d\mu = -\mathcal{S} H d\mu,$

(iii) $\dfrac{\partial}{\partial t} h_{ij} = \mathcal{L}_S h_{ij} + \dfrac{\partial^2 \mathcal{S}}{\partial h_q^p \partial h_l^k} \nabla_i h_q^p \nabla_j h_l^k + \dfrac{\partial \mathcal{S}}{\partial h_l^k} h_m^k h_l^m h_{ij} - (\beta + 1) h_{ik} h_j^k \mathcal{S},$

(iv) $\dfrac{\partial}{\partial t} h_j^i = \mathcal{L}_S h_j^i + \dfrac{\partial^2 \mathcal{S}}{\partial h_q^p \partial h_l^k} \nabla^i h_q^p \nabla_j h_l^k + \dfrac{\partial \mathcal{S}}{\partial h_l^k} h_m^k h_l^m h_j^i - (\beta - 1) h_k^i h_j^k \mathcal{S},$

(v) $\dfrac{\partial}{\partial t} H = \mathcal{L}_S H + \dfrac{\partial^2 \mathcal{S}}{\partial h_q^p \partial h_l^k} \nabla^i h_q^p \nabla_i h_l^k + \dfrac{\partial \mathcal{S}}{\partial h_l^k} h_m^k h_l^m H - (\beta - 1) |A|^2 \mathcal{S},$

(vi) $\begin{aligned} \dfrac{\partial}{\partial t} |A|^2 = \ & \mathcal{L}_S |A|^2 - 2 \dfrac{\partial \mathcal{S}}{\partial h_j^i} \nabla^i h_k^k \nabla_j h_k^l + 2 \dfrac{\partial^2 \mathcal{S}}{\partial h_q^p \partial h_l^k} \nabla^i h_q^p \nabla_j h_l^k h_i^j \\ & + 2 \dfrac{\partial \mathcal{S}}{\partial h_l^k} h_m^k h_l^m |A|^2 - 2(\beta - 1)\, \mathcal{S}\, h_k^i h_j^k h_i^j, \end{aligned}$

(vii) $\dfrac{\partial}{\partial t} \mathcal{S} = \mathcal{L}_S \mathcal{S} + \dfrac{\partial \mathcal{S}}{\partial h_l^k} h_m^k h_l^m \mathcal{S}.$

From equation (ii) we immediately deduce

$$\frac{d}{dt}\text{area}\,(\mathcal{M}_t) = \frac{d}{dt}\int_{\mathcal{M}_t} d\mu = -\int_{\mathcal{M}_t} \mathcal{S} H\, d\mu. \tag{4.3}$$

This shows that, among all possible functions $\mathcal{S}$ on $\mathcal{M}_t$ having fixed L^2 norm, the mean curvature is the one such that the area of $\mathcal{M}_t$ decreases most rapidly. Such a property is expressed by saying that the mean curvature flow is the *gradient flow* of the area functional.

It is also interesting to look at formula (4.3) in the case of the inverse mean curvature flow, where $\mathcal{S} = -1/H$. Then $-\mathcal{S}H \equiv 1$ and we obtain that the derivative of the area is the area itself. This implies that, regardless of the shape of the surface, we have

$$\text{area}\,(\mathcal{M}_t) = \text{area}\,(\mathcal{M}_0)\, e^t,$$

as long as the flow exists.

From now on we restrict our attention to the mean curvature flow. A first question we want to address concerns the length of the time interval where a solution exists.

Theorem 4.4. *Let $F_0 \colon \mathcal{M} \to \mathbb{R}^{n+1}$ be a smooth immersion of a closed n-dimensional manifold $\mathcal{M}$. Then the mean curvature flow (1.1)–(1.2) has a unique smooth solution, which can be defined in a maximal time interval $[0, T)$, where $0 < T < +\infty$, and satisfies $\max_{\mathcal{M}_t} |A|^2 \to \infty$ as $t \uparrow T$.*

Proof. The local existence and uniqueness is a particular case of Theorem 4.1. The finiteness of T follows from the avoidance principle of Theorem 4.2, by taking as $\mathcal{N}_0$ a sphere enclosing $\mathcal{M}_0$. Since the sphere shrinks to a point in finite time, and the two evolving surfaces remain disjoint, the smooth evolution of $\mathcal{M}_0$ must terminate not later than the one of $\mathcal{N}_0$. Recalling the maximal time of existence of a sphere, we see that the smooth evolution of $\mathcal{M}_t$ cannot exist beyond time $T = R_0^2/n$, where R_0 is the diameter of $\mathcal{M}_0$. The property that the curvature $|A|^2$ blows up as $t \to T$ can be proved by a standard contradiction argument, which relies on a priori estimates on the derivatives of A. We give a sketch of the proof, referring the reader to [38, Theorem 8.1] for more details. Suppose that $|A|^2$ is uniformly bounded for $t \in [0, T)$. Then also H is uniformly bounded, and therefore the immersions $F(\cdot, t)$ tend to a limit as $t \to T$. We denote this limit by $F(\cdot, T)$; we want to show that it is a smooth immersion. We use the estimates of Proposition 2.3 in [42], which show that, since $|A|^2$ is bounded, each derivative $\nabla^m A$, with $m > 0$, is also uniformly bounded. This implies that the surfaces $\mathcal{M}_t$ converge to a limiting smooth surface $\mathcal{M}_T$. Such a surface coincides with $F(\mathcal{M}, T)$; it remains to show that $F(\cdot, T)$ is a smooth parametrization of $\mathcal{M}_T$, that is, that the tangent vectors $\{\partial F/\partial x_i\}$ remain linearly independent up to time T. This is ensured by a lemma on equivalent metrics which is due to Hamilton [33, Lemma 14.2]. Thus, $F(\cdot, T)$ is a smooth immersion and the local existence theorem can be used to find a solution for times greater than T, in contradiction with our assumptions. $\square$

We call the time T the *singular time* of the mean curvature flow $\mathcal{M}_t$, and we call the behaviour of $\mathcal{M}_t$ as $t \to T$ *formation of singularities*. The above proof may suggest that any closed surface shrinks to a point as the singular time is reached, as in the case of the sphere. The following example shows that this is not the case.

Example 4.5 (The standard neckpinch). Suppose that $\mathcal{M}_0$ looks like two large balls connected by a cylindrical part (neck) which is very thin, in such a way that the mean curvature in the neck is much larger than in the balls. Then one expects that the radius of the neck goes to zero in a short time while the balls move little from their original position. The existence of surfaces with this property was first proved rigorously by Grayson [32]; we present here a simple proof of this fact given by Ecker [21]. Let us consider the function

$$f(y, t) = |y|^2 - \left(n - \frac{1}{2}\right) y_{n+1}^2 + t, \qquad y \in \mathbb{R}^{n+1}, \ t \in \mathbb{R}.$$

For a given family of immersions $F \colon \mathcal{M} \times [0, T) \to \mathbb{R}^{n+1}$ evolving by mean curvature flow, we denote by Φ the function f evaluated on $\mathcal{M}_t$, that is

$$\Phi(p, t) = f(F(p, t), t), \qquad p \in \mathcal{M}, \ t \in [0, T).$$

Then we have

$$\frac{\partial}{\partial t}\Phi = Df \cdot \frac{\partial F}{\partial t} + \frac{\partial f}{\partial t}$$
$$= -2HF \cdot \nu + (2n - 1)Hy_{n+1}\nu_{n+1} + 1.$$

Using (2.5) we find that $\Delta_{\mathcal{M}_t} y_{n+1}^2 = 2 - 2Hy_{n+1}\nu_{n+1} - 2\nu_{n+1}^2$. Hence, recalling also (2.7), we obtain

$$\Delta_{\mathcal{M}_t}\Phi = 2n - 2HF \cdot \nu - (2n - 1)(1 - Hy_{n+1}\nu_{n+1} - \nu_{n+1}^2).$$

It follows that

$$\frac{\partial}{\partial t}\Phi - \Delta_{\mathcal{M}_t}\Phi = -(2n - 1)\nu_{n+1}^2 \leq 0.$$

Therefore, if $\mathcal{M}_t$ is closed, the maximum principle implies that $\max_{\mathcal{M}_t} \Phi$ is non-increasing. Given $y \in \mathbb{R}^{n+1}$, we set $y = (\hat{y}, y_{n+1})$ with $\hat{y} \in \mathbb{R}^n$. The surface $\mathcal{H}$ of equation

$$|\hat{y}|^2 = \left(n - \frac{3}{2}\right)y_{n+1}^2 + 1$$

is a hyperboloid of rotation around the y_{n+1} axis. We now take any closed hypersurface $\mathcal{M}_0$ with the following properties:

- It is contained inside the hyperboloid $\mathcal{H}$; equivalently, it satisfies $f(p, 0) \leq 1$ for any $p \in \mathcal{M}_0$.

- It encloses two spheres Σ^1, Σ^2 of radius $2\sqrt{n}$ centered at $(0, L)$, $(0, -L)$ respectively, where L is chosen large enough so that this requirement is compatible with the former one.

Let us now consider the mean curvature flow $\mathcal{M}_t$ starting from $\mathcal{M}_0$. By the avoidance principle (Theorem 4.2), the region enclosed by $\mathcal{M}_t$ must contain the two spheres Σ_t^1, Σ_t^2 which are the evolution of Σ^1, Σ^2. We recall that, by Example 3.1, Σ_t^1, Σ_t^2 are spheres centered at $(0, \pm L)$ and exist for all $t < 2$. On the other hand, since $f(\cdot, 0) \leq 1$ on $\mathcal{M}_0$, the maximum principle argument above implies that $f(\cdot, t) \leq 1$ on $\mathcal{M}_t$. Observe that for $t = 1$ the region where $f(y, t) \leq 1$ is a cone with vertex at the origin, and that the points $(0, \pm L)$ belong to distinct connected components of the interior of the cone. This shows that the singular time T of the flow $\mathcal{M}_t$ satisfies $T \leq 1$, and that $\mathcal{M}_t$ does not shrink to a point as $t \to T$ since it encloses two balls with positive radius. Such a construction can be generalized to other flows of the form (1.4), see [2]. $\qquad\square$

The above example has motivated the search for suitable weak solutions to the mean curvature flow, in order to define a generalized evolution even after time T. In a physical model, for instance, the interface continues its evolution after time T, although in a non-smooth way; we intuitively expect that it splits in two parts, each of them flowing independently. For topological applications, it is also important to continue the flow until we have enough information on all parts of the surface to describe its global structure.

As we have recalled in the introduction, many authors in the past have introduced weak solutions to the mean curvature flow. At the end of these notes we will present a more recent approach, due to Huisken and the author [48], which is based on a surgery procedure, following Hamilton's program for the Ricci flow. This approach is more suitable for geometric applications because it allows us to keep track of the changes of topology of the evolving surface. Let us mention that the situation is different in the case of the Ricci flow, because no notion of weak solution is known there, and the flow with surgeries by Hamilton and Perelman is the only available definition of a flow past the singularities.

In contrast to the case of closed surfaces which always develop singularities, the flows of entire graphs by mean curvature exist for all times, as shown by the next result by Ecker and Huisken [22]:

Theorem 4.6. *Let $\mathcal{M}_0$ be the graph of a function $x_{n+1} = u(x_1, \ldots, x_n)$ defined for all $(x_1, \ldots, x_n) \in \mathbb{R}^n$. Then there exists a solution $\mathcal{M}_t$ of the mean curvature flow with initial data $\mathcal{M}_0$ which is defined for all $t \in (0, +\infty)$ and is a graph over $\mathbb{R}^n$ for all t.*

An interesting feature of this result is that, in contrast to other parabolic equations (e.g., the linear heat equation), no growth restriction on the function u is necessary to obtain existence of solutions. This is related to the geometric interpretation of the equation: in fact, a fast growth of u does not imply a fast growth of the curvature of the graph of u. On the other hand, the uniqueness of the solution obtained in the above theorem is still an open problem (we are not in the framework of Theorem 4.1 because the surfaces here are non-compact).

5 Invariance properties

A first step in the analysis of singularities is to observe that certain significant geometric properties are invariant under the flow, as a consequence of the maximum principle. We collect in this section some elementary, but significant, examples. Let us first rewrite some of the evolution equations of Lemma 4.3 in the case of the mean curvature flow $\mathcal{S} = H$.

Lemma 5.1. *If $\mathcal{M}_t$ evolves by mean curvature flow, then we have*

(i) $\dfrac{\partial}{\partial t} h^i_j = \Delta h^i_j + |A|^2 h^i_j,$

(ii) $\dfrac{\partial}{\partial t} H = \Delta H + |A|^2 H,$

(iii) $\dfrac{\partial}{\partial t} |A|^2 = \Delta |A|^2 - 2|\nabla A|^2 + 2|A|^4.$

We can immediately deduce some first examples of properties which are invariant under the flow.

Proposition 5.2. *Let $\mathcal{M}_t$, $t \in [0, T)$, be a family of closed surfaces evolving by mean curvature flow.*

(i) *If $H \geq 0$ on $\mathcal{M}_0$, then $H > 0$ on $\mathcal{M}_t$ for any $t \in (0, T)$.*

(ii) *If $|A|^2 \leq cH^2$ on $\mathcal{M}_0$, then $|A|^2 \leq cH^2$ on $\mathcal{M}_t$ for any $t \in (0, T)$.*

Proof. Part (i) follows from Lemma 5.1 and the strong maximum principle. To obtain (ii), we compute the evolution equation of $f = |A|^2/H^2$. We obtain, by Lemma 5.1 and a straightforward computation,

$$\frac{\partial f}{\partial t} = \Delta f + \frac{2}{H} \langle \nabla H, \nabla f \rangle - \frac{2}{H^4} |H \nabla_i h_{kl} - \nabla_i H \, h_{kl}|^2. \tag{5.1}$$

Thus, the maximum principle implies that the maximum of f is non-increasing. $\qquad\square$

Corollary 5.3. *Let $\mathcal{M}_t$, $t \in [0, T)$, be a family of closed surfaces evolving by mean curvature flow.*

(i) *If $H > 0$ on $\mathcal{M}_0$, then there is an $\varepsilon_0 > 0$ such that $\varepsilon_0 |A|^2 \leq H^2 \leq n|A|^2$ everywhere on $\mathcal{M}_t$ for all $t \in (0, T)$.*

(ii) *If $\mathcal{M}_0$ has positive scalar curvature, the same holds for $\mathcal{M}_t$ for all $t \in (0, T)$.*

Proof. By the compactness of $\mathcal{M}_0$, if $H > 0$ everywhere then we also have $H^2 \geq \varepsilon_0 |A|^2$ everywhere for some $\varepsilon_0 > 0$. We can apply Proposition 5.2(ii) to obtain the first inequality in (i). Inequality $H^2 \leq n|A|^2$ follows instead from the algebraic identity (2.3). Part (ii) is also a consequence of Proposition 5.2(ii) because, by (2.4), positive scalar curvature is equivalent to $H^2/|A|^2 > 1$. $\qquad\square$

Corollary 5.3(ii) is a particular case of a more general property of the elementary symmetric polynomials of the curvatures, as we now proceed to show. We recall that the *elementary symmetric polynomial* of degree k in n variables $\lambda_1, \ldots, \lambda_n$ is defined as

$$S_k = \sum_{1 \leq i_1 < i_2 < \cdots < i_k \leq n} \lambda_{i_1} \lambda_{i_2} \ldots \lambda_{i_k}$$

for $k = 1, \ldots, n$. In particular, $S_1 = H$, S_2 is half of the scalar curvature and S_n is the Gauss curvature. It is not difficult to show that

$$\lambda_1 \geq 0, \ldots, \lambda_n \geq 0 \quad \Longleftrightarrow \quad S_1 \geq 0, \ldots, S_n \geq 0. \tag{5.2}$$

These polynomials enjoy various concavity properties. The interesting one for our purposes is the following [51, 47].

Theorem 5.4. *Let $\Gamma_k \subset \mathbb{R}^n$ denote the connected component of the set*

$$\{\lambda \in \mathbb{R}^n : S_k(\lambda) > 0\}$$

which contains the positive cone. Then $S_l > 0$ on Γ_k for all $l = 1, \ldots, k$ and the quotient S_{k+1}/S_k is concave on Γ_k.

The above properties remain unchanged if we regard the polynomials S_k as functions of the Weingarten operator, instead of the principal curvatures, because we have the following result; see [6, Lemma 2.2] or [47, Lemma 2.11].

Theorem 5.5. *Let $f(\lambda_1, \ldots, \lambda_n)$ be a symmetric convex (concave) function and let $F(A) = f(\text{eigenvalues of } A)$ for any $n \times n$ symmetric matrix A whose eigenvalues belong to the domain of f. Then F is convex (concave).*

The concavity of the above expressions allows us to apply the maximum principle to obtain invariance properties. This will be clear after deriving the following evolution equation.

Proposition 5.6. *Let $F(h^i_j)$ be a function homogeneous of degree 1. Let $\mathcal{M}_t$ be a family of closed surfaces evolving by mean curvature flow such that h^i_j belongs to the domain of F everywhere. Then*

$$\frac{\partial}{\partial t}\frac{F}{H} - \Delta\frac{F}{H} = \frac{2}{H}\left\langle \nabla H, \nabla\frac{F}{H} \right\rangle - \frac{1}{H}\frac{\partial^2 F}{\partial h^i_j \partial h^k_l}\nabla^p h^j_i \nabla_p h^k_l.$$

As a consequence, if F is concave (convex), any estimate of the form $F \geq cH$ (resp. $F \leq cH$) is preserved.

Proof. A straightforward computation, using Lemma 5.1(iii)-(iv) and Euler's theorem on homogeneous functions, yields

$$\frac{\partial}{\partial t}\frac{F}{H} = \frac{1}{H}\frac{\partial F}{\partial h^i_j}(\Delta h^i_j + |A|^2 h^i_j) - \frac{F}{H^2}(\Delta H + |A|^2 H)$$

$$= \Delta\frac{F}{H} + \frac{2}{H}\left\langle \nabla H, \nabla\frac{F}{H} \right\rangle - \frac{1}{H}\frac{\partial^2 F}{\partial h^i_j \partial h^k_l}\nabla^p h^j_i \nabla_p h^k_l.$$

$\square$

In particular, the previous proposition can be applied to $F = S_{k+1}/S_k$, provided that $S_k \neq 0$. This leads to the following result, which generalizes Corollary 5.3.

Proposition 5.7. *Let $\mathcal{M}_0$ be a closed hypersurface such that $S_k > 0$ everywhere for a given $k \in \{1, \ldots, n\}$ and let $\mathcal{M}_t$ be its evolution by mean curvature flow. Then, for any $l = 2, \ldots, k$ there exists γ_l such that $S_l \geq \gamma_l H^l > 0$ on $\mathcal{M}_t$ for all $t \in (0, T)$.*

Proof. We first observe that on $\mathcal{M}_0$ the curvatures $(\lambda_1, \ldots, \lambda_n)$ belong everywhere to the set Γ_k defined in Theorem 5.4. In fact, by hypothesis they are contained in a connected component of the set where $S_k > 0$. In addition, the connected component must be the one that contains the positive cone because every closed surface has at least one point where all curvatures are positive.

By Theorem 5.4, we have $S_l > 0$ on $\mathcal{M}_0$ for $l = 1, \ldots, k$ and so, by compactness, we have $S_l \geq c_l H S_{l-1}$ for suitable constants $c_l > 0$, for any $l = 2, \ldots, k$. We know from Proposition 5.2 that $H > 0$ everywhere on $\mathcal{M}_t$ for $t \in (0, T]$. Then we can consider the quotient $S_2 / H^2 = S_2 / S_1 H$. It is defined for every t, it is greater than c_2 at time zero, and its minimum is non-decreasing by Proposition 5.6. It follows that $S_2 \geq c_2 H^2$ for $t \in (0, T)$.

We now apply the same procedure to the quotient $S_3 / S_2 H$ to conclude that it is greater than c_3 for $t \in (0, T)$, i.e., $S_3 \geq c_3 S_2 H \geq c_3 c_2 H^3$. After finitely many iterations, we obtain the conclusion. $\qquad\square$

Further invariance properties for the mean curvature flow can be obtained using Hamilton's maximum principle for tensors [34, §4]. Let us first recall a definition. We say that an immersed surface $\mathcal{M}$ is *k-convex*, for some $1 \leq k \leq n$, if the sum of the k smallest curvatures is non-negative at every point of $\mathcal{M}$. In particular, 1-convexity coincides with convexity, while n-convexity means non-negativity of the mean curvature H. Then we have the following result.

Proposition 5.8. *If a closed hypersurface $\mathcal{M}_0$ satisfies $\lambda_1 + \cdots + \lambda_k \geq \alpha H$ for some $\alpha \geq 0$ and $1 \leq k \leq n$, then the same holds for its evolution by mean curvature flow $\mathcal{M}_t$. In particular, if $\mathcal{M}_0$ is k-convex, then so is $\mathcal{M}_t$.*

Proof. The result follows from Hamilton's maximum principle for tensors, provided we show that the inequality $\lambda_1 + \cdots + \lambda_k \geq \alpha H$ describes a convex cone in the set of all matrices, and that this cone is invariant under the system of ordinary differential equations $dh_j^i / dt = |A|^2 h_j^i$, which is obtained by dropping the diffusion term in the evolution equation for the Weingarten operator in Lemma 5.1.

If we denote by $W(v_1, v_2)$ the Weingarten operator applied to two tangent vectors v_1, v_2 at any point, we have

$$\lambda_1 + \cdots + \lambda_k = \min\{W(e_1, e_1) + \cdots + W(e_k, e_k) :$$
$$\langle e_i, e_j \rangle = \delta_{ij} \text{ for all } 1 \leq i \leq j \leq k\}.$$

This shows that $\lambda_1 + \cdots + \lambda_k$ is a concave function of the Weingarten operator, being the infimum of a family of linear maps. Therefore the inequality $\lambda_1 + \cdots + \lambda_k \geq \alpha H$ describes a convex cone of matrices. In addition, the system $dh_j^i / dt = |A|^2 h_j^i$ changes the Weingarten operator by homotheties, and thus leaves any cone invariant. The conclusion follows. $\qquad\square$

In particular, we obtain that convex surfaces remain convex under the flow. Observe that the same property also follows from Proposition 5.7 by taking $k = n$ and taking into account property (5.2).

6 Singular behaviour of convex surfaces

In the next two sections we shall see some results showing that, roughly speaking, the convexity properties of a surface evolving by mean curvature flow improve when a singularity is formed. We begin with the case of convex surfaces.

Theorem 6.1. *Let $\mathcal{M}_0$ be an n-dimensional closed convex surface embedded in $\mathbb{R}^{n+1}$. Then $\mathcal{M}_t$ shrinks to a point as $t \to T$. In addition, if we choose a suitable rescaling factor $\rho(t)$, then the surfaces $\rho(t)\mathcal{M}_t$ converge to a sphere as $t \to T$.*

The behaviour described in the above theorem is usually summarized by saying that the surfaces converge to a "round point". This result was proved by Huisken in [38] in the case $n \geq 2$ and by Gage and Hamilton [26] when $n = 1$. We describe here some of the main ideas in the proof in the case $n > 1$, since they play an important role in the later developments of the theory. Let us set

$$f = \frac{|A|^2}{H^2} - \frac{1}{n}.$$

Then equation (2.3) can be rewritten as

$$fH^2 = \sum_{i<j}(\lambda_i - \lambda_j)^2.$$

Thus, f is non-negative and it measures how much the curvatures differ from each other. It vanishes identically on a surface if and only if the surface is a sphere. This approach is suggested by [33, §8], where Hamilton considered a similar function of the eigenvalues of the Ricci tensor.

We have seen in Proposition 5.2 that the maximum of f is non-increasing. To prove convergence to a sphere one needs some stronger estimate, showing that f tends to zero as the singular time is approached. To this purpose, Huisken considered the function $f_\sigma = fH^\sigma$ for a suitably small $\sigma > 0$. Observe that f_σ is a homogeneous function of the curvatures of degree $\sigma > 0$; thus, one would expect f_σ to blow up as the singular time T is approached. The next theorem shows instead that it remains bounded, and this is one of the crucial steps in the proof of Theorem 6.1.

Theorem 6.2. *If $\sigma > 0$ is small enough, then the function f_σ is uniformly bounded for $t \in [0, T)$.*

Proof. Let us first remark that a similar result holds for the analogous function considered in [33] for the Ricci flow. However, the method of proof is quite different. In fact, the result of [33] follows from an application of the maximum principle. In our case, instead, the additional factor H^σ induces the presence of a positive zero-order term in the evolution equation for f_σ that cannot be directly compensated by the other terms. More precisely, one finds

$$\frac{\partial f_\sigma}{\partial t} \leq \Delta f_\sigma + \frac{2(1-\sigma)}{H}\langle \nabla H, \nabla f_\sigma \rangle - \frac{2}{H^{4-\sigma}}|H\nabla_i h_{kl} - h_{kl}\nabla_i H| + \sigma|A|^2 f_\sigma. \quad (6.1)$$

Thus, a more elaborate procedure is needed to estimate f_σ. Let us first state a useful lower bound for the gradient term in the above inequality. One can prove that on convex surfaces (and in fact under more general hypotheses) there exists c such that

$$|H\nabla_i h_{kl} - h_{kl}\nabla_i H| \geq cH^2|\nabla H|^2 \tag{6.2}$$

(see [38, Lemma 2.3]). We now integrate the inequality on the manifold and try to estimate the L^p norm of f_σ. After integrating by parts we obtain

$$\frac{d}{dt} \int_{\mathcal{M}_t} f_\sigma^p \, d\mu \leq -\frac{p(p-1)}{2} \int_{\mathcal{M}_t} f_\sigma^{p-2}|\nabla f_\sigma|^2 \, d\mu - \frac{p}{c} \int_{\mathcal{M}_t} \frac{f_\sigma^{p-1}}{H^{2-\sigma}}|\nabla H|^2 \, d\mu$$

$$+ p\sigma \int_{\mathcal{M}_t} |A|^2 f_\sigma^p \, d\mu. \tag{6.3}$$

To show that the last term can be compensated by the other two, we need some estimate involving both zero-order curvature terms and gradient terms. To this purpose, we recall the identity [38, Lemma 2.1]

$$\frac{1}{2}\Delta|A|^2 = \langle h_{ij}, \nabla_i \nabla_j H \rangle + |\nabla A|^2 + Z, \tag{6.4}$$

where $Z = H\sum \lambda_i^3 - \left(\sum \lambda_i^2\right)^2$. Using this equality, we can compute

$$\Delta f_\sigma \geq \frac{2}{H^{2-\sigma}} h^{ij}\nabla_i\nabla_j H + \frac{2}{H^{2-\sigma}} Z$$

$$- \frac{2(1-\sigma)}{H} \langle \nabla H, \nabla f_\sigma \rangle + \left(\frac{2}{nH^{1-\sigma}} - \frac{2-\sigma}{H} f_\sigma\right)\Delta H.$$

After integrating this inequality on $\mathcal{M}_t$ and performing some standard computations we obtain that, for all $\eta > 0$,

$$\int \frac{1}{H^{2-\sigma}} f_\sigma^{p-1} Z \, d\mu \leq (2\eta p + 5) \int \frac{1}{H^{2-\sigma}} f_\sigma^{p-1}|\nabla H|^2 \, d\mu$$

$$+ \eta^{-1}(p-1) \int f_\sigma^{p-2}|\nabla f_\sigma|^2 \, d\mu. \tag{6.5}$$

On the other hand, it can be shown [38, Lemma 2.3(i)] that on a uniformly convex surface, say $h_{ij} \geq \varepsilon H g_{ij}$, we have

$$Z \geq n\varepsilon^2 H^{4-\sigma} f_\sigma > n\varepsilon^2 |A|^2 H^{2-\sigma} f_\sigma. \tag{6.6}$$

Thus, we can combine estimates (6.3), (6.5) and (6.6) and choose η appropriately to show that, for p suitably large and for σ suitably small the L^p norm of f_σ is decreasing in time.

This property is the starting point for a Stampacchia iteration procedure to obtain that the L^∞ norm of f_σ is bounded. The proof also relies on a Sobolev inequality for immersed manifolds due to Michael and Simon [53]. For the details, see [38, §5]. □

Several steps remain to complete the proof of Theorem 6.1. Roughly speaking, the above result shows that, at the points where the curvature becomes unbounded, the Weingarten operator approaches the one of a sphere. One then needs to show that the curvature becomes unbounded in the whole surface when the singular time is approached. The main steps are a gradient estimate for the mean curvature and an application of Myers' theorem; see [38].

It is interesting to remark that the above proof does not apply to the one-dimensional case. The proof of Gage and Hamilton in [26] for curves in the plane was obtained by a different method. Actually, in that case the result can be generalized to any embedded curve, as it was shown by Grayson [30].

Theorem 6.3. *Let Γ_0 be a closed embedded curve in the plane and let Γ_t, with $t \in [0, T)$, be its evolution by mean curvature flow. Then there exists $t_0 < T$ such that Γ_t is convex for all $t \geq t_0$. As a consequence, the result of [26] applies and the curve (up to a rescaling) converges to a circle as $t \to T$.*

Other types of singular behaviour arise if one considers immersed plane curves with self-intersections. Let us mention for instance the result by Angenent [9] dealing with the case of a convex immersed curve looking like a cardioid; that is, with one small loop enclosed by the rest of the curve. Then at the singular time the loop shrinks to a point forming a cusp, while the rest of the curve is still non-singular. After rescaling near the singularity (by a procedure similar to the one that we describe later in Section 8) the asymptotic profile of the curve is given by the "grim reaper" curve of Example 3.2.

In the remainder of the section, we give a brief survey of the many extensions of Huisken's theorem which have been obtained in the following years. Let us first mention some results, again due to Huisken, about hypersurfaces immersed in a Riemannian manifold. In [39] it was proved that if the principal curvatures of the initial surface satisfy a suitable lower bound, depending on the properties of the ambient manifold, then there is again convergence to a round point. The paper [40] considered the case where the ambient manifold is a sphere S^{n+1}. In this case there is a class of initial surfaces for which an alternative holds: either there is convergence to a round point in finite time, or there is convergence in infinite time to a smooth totally geodesic submanifold of S^{n+1}, roughly speaking, to a "big S^n". A result in the same spirit was proved by Grayson [31] for curves embedded in a Riemannian two-dimensional manifold; under suitable assumptions on the ambient space, he proved that the curve either converges to a round point or to a geodesic.

Coming back to the case of a euclidean ambient space, there is a wide literature concerning the extension of Theorem 6.1 to general curvature flows of the form (1.4). For certain flows, the convergence to a round point has been proved for any convex surface. In other cases, the result is known under the assumption that the initial surface satisfies a pinching condition of the form $\lambda_n \leq C\lambda_1$ at every point, where $C > 1$ is a suitable constant depending on the flow and on the dimension. Among the various situations considered, the ones where the strongest results are known are usually those where the speed is a function homogeneous of degree 1 in the curvatures, and those where the dimension of the evolving surfaces is $n = 2$. We summarize some of the main results below.

We mention first the case of the flow driven by the Gauss curvature or, more generally, by the power β of the Gauss curvature, for some $\beta > 0$. Some partial results were obtained in [63], a paper which has been influential in the following developments of the subject. The convergence to a round point of convex surfaces has been proved in the case $\beta = 1/n$ in [15], and when $\beta = 1$ and $n = 2$ in [7]. For a power $\beta > 1/n$, in general dimension, Huisken's result is known to hold if the initial value satisfies a pinching condition [15]. If $\beta < 1/n$, the result may fail; in fact, there are examples of convex homothetically shrinking surfaces, which are not spherical; see [8]. The convergence to a round point under a pinching condition has been proved also when the speed is a power of the scalar curvature [2, 16] or a power greater than one of the mean curvature [59].

The validity of Huisken's result, without pinching conditions, was extended to a large class of flows with one-homogeneous speed in [6]. More recently, the same has been proved for various significant speeds with higher homogeneity degree in [57] in the two-dimensional case.

There are other flows where we observe convergence of convex surfaces to a spherical profile in infinite time. One example is the inverse mean curvature flow, where $\mathcal{S} = -1/H$. It was proved in [64] that convexity is preserved, that convex solutions exist for all times and converge to a sphere as $t \to \infty$ after rescaling. Let us also mention the so-called volume preserving curvature flows, which have the form

$$\frac{\partial F}{\partial t}(p, t) = (-\mathcal{S}(p, t) + h(t)) \nu(p, t),$$

where $h(t)$ is the mean value of $\mathcal{S}$ on $\mathcal{M}_t$. Thanks to the $h(t)$ term, the volume enclosed by $\mathcal{M}_t$ remains constant. There are various cases where it is known that convex solutions exist for all times and converge to a sphere as $t \to \infty$. This was first proved in [41] when the speed $\mathcal{S}$ is the mean curvature flow. Among the other results, let us quote the ones of [52], where a large class of functions $\mathcal{S}$ homogeneous of degree 1 is considered, and the paper [13], which studies the case of the volume preserving mean curvature flow in hyperbolic spaces.

7   Convexity estimates

We now consider the formation of singularities for the so-called *mean convex* hypersurfaces, that is, the ones with non-negative mean curvature everywhere. As we have seen in Theorem 5.2, this property is preserved by the mean curvature flow. For the study of singularities, mean convexity is a significant generalization of convexity. For instance, it is enough in general to allow for the neckpinch behaviour described in Example 4.5; therefore, mean convex surfaces do not necessarily shrink to a point at the singular time. An important result in the analysis of these surfaces is the following estimate on the elementary symmetric polynomials of the curvatures, proved in [47].

Theorem 7.1. *Let $\mathcal{M}_0 \subset \mathbb{R}^{n+1}$ be a closed mean convex immersed hypersurface and let $\mathcal{M}_t$, $t \in [0, T)$, be its evolution by mean curvature flow. Then, for any $\eta > 0$ there exists $C = C(\eta, \mathcal{M}_0)$ such that $S_k \geq -\eta H^k - C$ for any $k = 2, \ldots, n$ on $\mathcal{M}_t$ for any $t \in [0, T)$.*

Such an estimate easily implies the following one, which has a more immediate interpretation.

Theorem 7.2. *Under the same hypotheses of the previous theorem, for any $\eta > 0$ there exists $C = C(\eta, \mathcal{M}_0)$ such that $\lambda_1 \geq -\eta H - C$ on $\mathcal{M}_t$ for any $t \in [0, T)$.*

The interest of the above estimate lies in the fact that η can be chosen arbitrarily small and C is a constant which does not depend on the curvatures. Thus we see that, roughly speaking, the negative curvatures become negligible with respect to the others when the singular time is approached. This implies that the surface becomes asymptotically convex near a singularity. For this reason we call the estimates of the theorems above *convexity estimates*. Let us observe that the result of Theorem 7.2 is very similar to a well-known estimate in the Ricci flow, usually called *Hamilton–Ivey estimate* [36, Theorem 24.4]. In contrast to our result, Hamilton–Ivey does not require curvature assumptions, but holds only in the three-dimensional case. Observe that Theorem 7.2 cannot be valid without any curvature assumption, even in low dimensions. In fact, Angenent [10] has proved the existence of a torus which shrinks homothetically under mean curvature flow, a behaviour which is incompatible with the validity of the convexity estimates.

Sketch of the proof of Theorem 7.1. We only describe the main idea of the proof. The result is obtained by an induction procedure. The estimate on S_1 follows from the assumptions, because $S_1 = H$. We then suppose that we have proved the theorem for the polynomials $S_1, \ldots, S_k$ and we want to prove the estimate for S_{k+1}.

The basic idea is to consider the quotient $Q_{k+1} = S_{k+1}/S_k$. We have seen in Proposition 5.6 that the function Q_{k+1}/H is bounded from below. In analogy with the proof of Theorem 6.2, we can try to prove that the same property holds for $Q_{k+1}/H^{1-\sigma}$, if $\sigma > 0$ is small enough. In this way we could show that the negative part of S_{k+1} is of lower order near a singularity and obtain the desired estimate.

The above argument cannot be applied in this form, because in general S_k is not positive everywhere and therefore the quotient Q_{k+1} is not even well defined. However, our induction hypothesis shows that S_k is "almost" non-negative. This allows us to introduce a suitable perturbation of the second fundamental form such that the corresponding perturbed S_k is non-negative. For the sake of simplicity, we will neglect the perturbation in the following computations. We define the function

$$f = f_{\sigma,\eta} = \frac{-Q_{k+1} - \eta H}{H^{1-\sigma}}$$

where $\sigma, \eta > 0$. The reason for introducing the additional term ηH will be explained later. A straightforward calculation yields the evolution equation

$$\frac{\partial f}{\partial t} = \Delta f + \frac{2(1-\sigma)}{H} \langle \nabla H, \nabla f \rangle - \frac{\sigma(1-\sigma)}{H^2} f |\nabla H|^2$$
$$+ \frac{1}{H^{1-\sigma}} \frac{\partial^2 Q_{k+1}}{\partial h_j^i \partial h_q^p} \nabla^m h_j^i \nabla_m h_q^p + \sigma |A|^2 f. \tag{7.1}$$

The presence of the reaction term due to σ does not allow us to estimate f from above directly using the maximum principle. Therefore, we use a strategy similar to the one of [38]. The crucial step consists of deriving L^p estimates on the positive part of f for large p. We observe that, at the points where $f \geq 0$, we have $S_{k+1} \leq -\eta H$, i.e., S_{k+1} is strictly negative. This is useful for our estimates, because the concavity properties of Q_{k+1} are worse at the points where $Q_{k+1} = 0$. More precisely, it is possible to prove [47, Theorems 2.5 and 2.14] that for any $\eta > 0$ there exists $c > 0$ such that

$$\frac{\partial^2 Q_{k+1}}{\partial h_j^i \partial h_q^p} \nabla^m h_j^i \nabla_m h_q^p \leq -c \frac{|\nabla A|^2}{|A|}$$

at all points where $Q_{k+1} < -\eta H$. This gives a good gradient term in equation (7.1). On the other hand, it is possible to show that this good term can compensate the reaction term $\sigma |A|^2 f$ if they are both integrated on the part of the surface where $f \geq 0$. In this way we can prove [47, Proposition 3.6] that, for any $\eta > 0$ and p large enough, there exists $\sigma > 0$ such that the L^p norm of $(f_{\sigma,\eta})_+$ is non-increasing. This allows us to apply the same iteration procedure as in the proof of Theorem 6.2 to conclude that $f_{\sigma,\eta}$ is bounded from above for a suitable $\sigma > 0$. By definition of $f_{\sigma,\eta}$, such a bound implies that

$$S_{k+1} \geq S_k(-\eta H - CH^{1-\sigma}) \geq -2\eta H^{k+1} - C'$$

for a constant $C' = C'(C, \sigma, \eta)$. Since $\eta > 0$ is arbitrary, this proves Theorem 7.1. $\qquad\square$

8   Rescaling near a singularity

To study the singular behaviour of hypersurfaces evolving by mean curvature flow, we use a standard technique in partial differential equations which consists of rescaling the solutions near a singularity. Such an approach is often related to the invariance of the equation with respect to certain transformations. Roughly speaking, one proves that suitable rescalings of the flow converge to a smooth non-trivial limit, which describes the asymptotic profile of the surface near a singularity.

Before describing the rescaling procedure in detail, let us estimate the blowup rate of the curvature at the singular time. We note that the function $v(t) = 1/2(T-t)$, for $t \in [0,T)$, is a solution of the equation $v' = 2v^2$ and satisfies $v(t) \to +\infty$ as $t \to T$. Recalling Lemma 5.1(iii), a standard application of the maximum principle yields

$$\max_{\mathcal{M}_t} |A|^2 > \frac{1}{2(T-t)}. \tag{8.1}$$

This suggests a classification of singularities depending on whether the curvature satisfies an upper bound analogous to (8.1).

Definition 8.1. We say that a family of surfaces $\mathcal{M}_t$ evolving by mean curvature flow develops a singularity of *type I* if there is a $C \geq 1/2$ such that

$$\max_{\mathcal{M}_t} |A|^2 \leq \frac{C}{T-t}. \tag{8.2}$$

Otherwise we say that the singularity is of *type II*.

It is easy to check that a sphere or a cylinder develops a singularity of type I. It was shown in [38] that the same is true for all convex surfaces. However, there are examples of singularities of type II. For instance, the immersed curve considered by Angenent in [9] develops a singularity of type II. In dimension higher than 1, type II singularities can occur also in the embedded case, as shown by the next example.

Example 8.2 (The degenerate neckpinch). For a given $\lambda > 0$, let us set

$$\phi_\lambda(x) = \sqrt{(1-x^2)(x^2+\lambda)}, \qquad -1 \leq x \leq 1.$$

For any $n \geq 2$, let $\mathcal{M}^\lambda$ be the n-dimensional surface in $\mathbb{R}^{n+1}$ obtained by rotation of the graph of ϕ_λ. The surface $\mathcal{M}^\lambda$ looks like a dumbbell, where the parameter λ measures the width of the central part. Then, it is possible to prove the following properties (see [3]):

(a) if λ is large enough, the surface $\mathcal{M}_t^\lambda$ eventually becomes convex and shrinks to a point in finite time;

(b) if λ is small enough, $\mathcal{M}_t^\lambda$ exhibits a neckpinch singularity as in Example 4.5;

(c) there exists at least one intermediate value of $\lambda > 0$ such that $\mathcal{M}_t^\lambda$ shrinks to a point in finite time, has positive mean curvature up to the singular time, but never becomes convex. The maximum of the curvature is attained at the two points where the surface meets the axis of rotation.

In addition, it can be proved that the singularity is of type I in cases (a), (b) and of type II in case (c). The behaviour in (c) is called *degenerate neckpinch* and was first conjectured by Hamilton for the Ricci flow [36, §3]. The rigorous construction described above for the mean curvature flow was first given in [3]. Intuitively speaking, it is a limiting case of the neckpinch where the cylinder in the middle and the balls on the sides shrink at the same time. One can also build the example in an asymmetric way, with only one of the two balls shrinking simultaneously with the neck, while the other one remains non-singular. A detailed description of the asymptotic profile for a class of rotationally symmetric surfaces exhibiting a degenerate neckpinch has been done in [11].

Following Hamilton [36, §16], we define the rescaling in a different way depending on the type of the singularity. For type I, we take any sequence of times $\{t_h\}$ such that $t_h \uparrow T$ and choose $p_h \in \mathcal{M}$ such that

$$|A|^2(p_h, t_h) = \max_{\mathcal{M}_{t_h}} |A|^2.$$

In case II we choose the sequence (p_h, t_h) in a more restrictive way. For any $h \geq 1$, we choose $t_h \in [0, T - 1/h]$, $p_h \in \mathcal{M}$ such that

$$|A|^2(p_h, t_h)\left(T - \frac{1}{h} - t_h\right) = \max_{\substack{t \leq T - 1/h \\ p \in \mathcal{M}}} |A|^2(p, t)\left(T - \frac{1}{h} - t\right). \tag{8.3}$$

For both types of singularities we then set

$$L_h = |A|^2(p_h, t_h), \quad \alpha_h = -L_h t_h, \quad \omega_h = L_h\left(T - t_h - \frac{1}{h}\right) \tag{8.4}$$

and define for any h the family F_h of rescaled immersions,

$$F_h(\cdot, \tau) = \sqrt{L_h}\left(F\left(\cdot, \frac{\tau}{L_h} + t_h\right) - F(p_h, t_h)\right), \quad \tau \in [\alpha_h, \omega_h]. \tag{8.5}$$

It is immediate to check that $\mathcal{M}_{h,\tau} = F_h(\mathcal{M}, \tau)$ evolves by mean curvature flow. Then we have the following properties.

Lemma 8.3. *We have, as $h \to \infty$,*

$$t_h \to T, \quad L_h \to \infty, \quad \alpha_h \to -\infty, \quad \omega_h \to \Omega,$$

where Ω is a finite positive value if the singularity is of type I *and $\Omega = +\infty$ if it is of type* II. *Moreover, for any T_0, T_1 such that $-\infty < T_0 < T_1 < \Omega$, the surfaces $\mathcal{M}_{h,\tau}$ have uniformly bounded curvature for $\tau \in [T_0, T_1]$ and for h large enough.*

Proof. See [46, Lemma 4.4]. □

A uniform bound on the curvature implies bounds on all derivatives of the curvature, by the estimates of [22]. Then, by standard techniques, it is possible to show that a subsequence of the flows $\mathcal{M}_{h,\tau}$ converges smoothly to a limit evolving surface $\widetilde{\mathcal{M}}_\tau$ defined for $\tau \in (-\infty, \Omega)$.

Since the original flow satisfies the estimates of Theorem 7.1, we deduce that the rescaled surfaces $\mathcal{M}_{h,\tau}$ satisfy

$$S_{k,h} > -\eta H_h^k - L_h^{-k/2} C_\eta.$$

Letting $h \to \infty$, we find that the limit flow satisfies $\widetilde{S}_k > -\eta \widetilde{H}^k$ for all $\eta > 0$, and therefore $\widetilde{S}_k \geq 0$. Recalling property (5.2) we conclude:

Corollary 8.4. *Let the flow $\widetilde{\mathcal{M}}_\tau$ be the limit of rescalings of a flow $\mathcal{M}_t$ of closed mean convex surfaces, obtained by the procedure described above. Then the surfaces $\widetilde{\mathcal{M}}_\tau$ are convex. In addition, the flow $\widetilde{\mathcal{M}}_\tau$ is defined for $\tau \in (-\infty, \Omega)$, where Ω is finite if the singularity of $\mathcal{M}_t$ is of type I, and $\Omega = +\infty$ if the singularity is of type II.*

The above corollary has been also obtained by White [68] by completely different techniques. His approach also applies to the rescalings of weak solutions after the singular time.

We point out that the rescaled surfaces satisfy uniform bounds on the curvature and not on the diameter. The convergence to $\widetilde{\mathcal{M}}_\tau$ must be interpreted as a locally uniform convergence, and the surfaces $\widetilde{\mathcal{M}}_\tau$ can be unbounded. For instance, they are clearly unbounded if the original surfaces $\mathcal{M}_t$ do not shrink to a point as $t \to T$. Let us also observe that the limiting surfaces are convex, but in general not strictly convex. For example, the typical limit that can be observed in a neckpinch singularity is the cylinder $S^{n-1} \times \mathbb{R}$. However, if the limiting surfaces are not strictly convex, then they necessarily split as the product of a flat factor and of a strictly convex one, as shown by the following result.

Theorem 8.5. *Let $\widetilde{\mathcal{M}}_\tau$ be as in the previous corollary. If the surfaces $\widetilde{\mathcal{M}}_\tau$ are not strictly convex, then (up to a rigid motion) they can be written as*

$$\widetilde{\mathcal{M}}_\tau = \Gamma_\tau^k \times \mathbb{R}^{n-k},$$

where $1 \leq k \leq n$ and Γ_τ^k is a family of strictly convex k-dimensional surfaces.

Proof. See [47, Theorem 4.1] and the references therein. $\qquad\square$

To investigate the possible structure of the strictly convex factor Γ_τ^k, we need some further tools which will be given in the next section.

9 Huisken's monotonicity formula

An important result in the analysis of the mean curvature flow is the following *monotonicity formula*, which was proved by Huisken in [42]. We recall here the proof of this result and discuss its consequences in the analysis of the singularities.

Theorem 9.1. *Let $\mathcal{M}_t = F(\mathcal{M}, t)$ be a family of closed immersed surfaces evolving by mean curvature flow, with $t \in (t_0, t_1)$. For any $\bar{t} \geq t_1$, let $\rho(y, t)$ be the n-dimensional backward heat kernel in $\mathbb{R}^{n+1}$ centered at $(0, \bar{t})$, that is,*

$$\rho(y, t) = \frac{1}{(4\pi(\bar{t} - t))^{n/2}} \exp\left(-\frac{|y|^2}{4(\bar{t} - t)}\right), \qquad y \in \mathbb{R}^{n+1}, \ t < \bar{t}.$$

Then we have

$$\frac{d}{dt} \int_{\mathcal{M}_t} \rho(y, t) \, d\mu = -\int_{\mathcal{M}_t} \rho(y, t) \left| \frac{F^\perp}{2(\bar{t} - t)} - H\nu \right|^2 d\mu, \tag{9.1}$$

where $F^\perp$ is the normal component of F. In particular, the integral of ρ on $\mathcal{M}_t$ is non-increasing in time.

Proof. For a fixed $\tau \neq 0$, let us introduce the vector field $V : \mathbb{R}^{n+1} \to \mathbb{R}^{n+1}$,

$$V = y \exp\left(-\frac{|y|^2}{4\tau}\right)$$

where $y = (y_1, \ldots, y_{n+1})$. We can compute the tangential divergence of V using (2.11). Keeping into account that

$$\frac{\partial V_\alpha}{\partial y_\beta} = \left(\delta_{\alpha\beta} - \frac{y_\alpha y_\beta}{2\tau}\right) \exp\left(-\frac{|y|^2}{4\tau}\right), \qquad \alpha, \beta \in \{1, \ldots, n+1\}$$

we find that

$$\begin{aligned}
\operatorname{div}^M V &= \left(n + 1 - \frac{|y|^2}{2\tau}\right) \exp\left(-\frac{|y|^2}{4\tau}\right) - \left(|\nu|^2 - \frac{\langle y, \nu \rangle^2}{2\tau}\right) \exp\left(-\frac{|y|^2}{4\tau}\right) \\
&= \left(n - \frac{|y^T|^2}{2\tau}\right) \exp\left(-\frac{|y|^2}{4\tau}\right),
\end{aligned}$$

where y^T denotes the tangential component of y. Therefore, the divergence theorem (2.10) implies

$$\int_{\mathcal{M}_t} \left(n - \frac{|y^T|^2}{2\tau}\right) \exp\left(-\frac{|y|^2}{4\tau}\right) d\mu = \int_{\mathcal{M}_t} H(V \cdot \nu) \, d\mu. \tag{9.2}$$

Let us observe that, for a general function $f(y, t)$, we have, by definition of the flow and by Lemma 4.3(ii),

$$\begin{aligned}
\frac{d}{dt} \int_{\mathcal{M}_t} f(y, t) \, d\mu &= \int_{\mathcal{M}_t} \left(\frac{\partial f}{\partial t} + Df \cdot \frac{\partial F}{\partial t}\right) d\mu + \int_{\mathcal{M}_t} f(y, t) \frac{\partial}{\partial t} \, d\mu \\
&= \int_{\mathcal{M}_t} \left(\frac{\partial f}{\partial t} - HDf \cdot \nu - H^2 f\right) d\mu. \tag{9.3}
\end{aligned}$$

We want to apply the above formula to the function ρ. Let us set from now on $\tau = \bar{t} - t$. We first compute

$$\frac{\partial \rho}{\partial t} = \left(\frac{n}{2\tau} - \frac{|y|^2}{4\tau^2} \right) \rho, \qquad D\rho = -\frac{y}{2\tau} \rho.$$

Therefore

$$\frac{\partial \rho}{\partial t} - HD\rho \cdot \nu - H^2 \rho = \frac{n}{2\tau} \rho - \left| \frac{y}{2\tau} - H\nu \right|^2 \rho + HD\rho \cdot \nu. \tag{9.4}$$

Observe in addition that $D\rho$ coincides with the field V defined before, up to a factor which is space-independent. Therefore (9.2) implies

$$\int_M H(D\rho \cdot \nu) \, d\mu = - \int_M \left(n - \frac{|y^T|^2}{2\tau} \right) \frac{\rho}{2\tau} \, d\mu. \tag{9.5}$$

Putting together (9.3), (9.4) and (9.5) we obtain

$$\frac{d}{dt} \int_{M_t} \rho(y,t) \, d\mu = \int_{M_t} \left(-\left| \frac{y}{2\tau} - H\nu \right|^2 \rho + \frac{|y^T|^2}{4\tau^2} \rho \right) d\mu. \tag{9.6}$$

Finally, observe that

$$\left| \frac{y}{2\tau} - H\nu \right|^2 = \left| \left(\frac{y^\perp}{2\tau} - H\nu \right) + \frac{y^T}{2\tau} \right|^2 = \left| \frac{y^\perp}{2\tau} - H\nu \right|^2 + \frac{|y^T|^2}{4\tau^2},$$

by Pythagoras' theorem. Plugging this in (9.6), and recalling that $F \equiv y$ on M, we obtain the conclusion. □

Clearly, in deriving a result such as the above one, the difficult part consists of guessing the correct statement. The idea of considering the heat kernel was suggested by another monotonicity formula which was known for semilinear heat equations in euclidean space; see the references in [42]. Also, it is interesting to observe that minimal surfaces (which are stationary solutions of the mean curvature flow), satisfy a formula where there is a monotonicity with respect to the radius of the sphere where a certain quantity is integrated (see [21, Appendix D]). A local version of the monotonicity formula for the mean curvature flow has been obtained afterwards by Ecker [20].

Finding monotone quantities in a partial differential equation often has useful applications. Well-known recent examples are contained in the work of Perelman [55] for the Ricci flow. In the case of the mean curvature flow the monotonicity formula has various consequences. In particular, it is useful for the classification of the blowups $\widetilde{\mathcal{M}}_\tau$ defined in the previous section, as the next result shows.

Theorem 9.2. *Let $\mathcal{M}_t$ have a singularity of type I as $t \to T$. Then, any flow $\widetilde{\mathcal{M}}_\tau$, with $\tau \in (-\infty, \Omega)$, obtained as a limit of rescalings by the procedure of the previous section, is a homothetically shrinking solution of the mean curvature flow. More precisely, $\widetilde{\mathcal{M}}_\tau$ (up to a translation) has the form*

$$\widetilde{\mathcal{M}}_\tau = \sqrt{\frac{\Omega - \tau}{\Omega}} \, \widetilde{\mathcal{M}}_0, \qquad \text{for all } \tau < \Omega, \tag{9.7}$$

where $\widetilde{\mathcal{M}}_0$ is the image of an immersion $\widetilde{F}_0$ which satisfies

$$\widetilde{F}_0 \cdot \nu = 2\Omega \, H. \tag{9.8}$$

Proof. We only give a sketch of the proof of this result, which can be found in [42] and is based on the monotonicity formula. Let us denote by

$$\widetilde{F} \colon \widetilde{\mathcal{M}} \times (-\infty, \Omega) \to \mathbb{R}^{n+1}$$

the immersion obtained as a limit of rescalings such that $\widetilde{F}(\widetilde{\mathcal{M}}, \tau) = \widetilde{\mathcal{M}}_\tau$. The main step of the argument, which we do not reproduce here, consists of showing that, roughly speaking, the right-hand side of (9.1), with $\bar{t} = \Omega$, must be identically zero on $\widetilde{\mathcal{M}}_\tau$. This is proved by a contradiction argument which uses the special properties of a limit of rescalings and the assumption that the singularity is of type I. By setting equal to zero the right-hand side of (9.1) for $\tau = 0$, one obtains property (9.8) above, where we have set $\widetilde{F}_0 = \widetilde{F}(\cdot, 0)$.

It is easy to see that (9.8) implies that the solution is homothetically shrinking. In fact, let us define the family of immersions

$$\Phi(p, t) = \sqrt{\frac{\Omega - \tau}{\Omega}} \, \widetilde{F}_0(p), \qquad p \in \widetilde{\mathcal{M}}, \ \tau < \Omega.$$

Clearly, $\Phi = \widetilde{F}$ for $\tau = 0$; we claim that the same is true for $\tau \neq 0$ up to a tangential diffeomorphism. To show this, observe first that the normal and the mean curvature at the point $\Phi(p, \tau)$ satisfy

$$\nu(p, \tau) = \nu(p, 0), \qquad H(p, \tau) = \sqrt{\frac{\Omega}{\Omega - \tau}} H(p, 0).$$

Then we can compute, using (9.8),

$$\frac{\partial \Phi}{\partial \tau}(p, \tau) \cdot \nu(p, \tau) = -\frac{\widetilde{F}_0(p)}{2\sqrt{\Omega(\Omega - \tau)}} \cdot \nu(p, \tau)$$

$$= -\sqrt{\frac{\Omega}{\Omega - \tau}} \frac{\widetilde{F}_0(p)}{2\Omega} \cdot \nu(p, 0)$$

$$= -\sqrt{\frac{\Omega}{\Omega - \tau}} H(p, 0) = -H(p, \tau).$$

Thus, the normal component of the speed of Φ is equal to the mean curvature. Observe that Φ is not necessarily a mean curvature flow because the speed may have a tangential component as well. However, we can compose Φ with a tangential diffeomorphism and obtain a family of immersions evolving by mean curvature flow. By uniqueness, this family coincides with $\widetilde{F}$. Since a tangential diffeomorphism leaves the image of the immersion unchanged, the surfaces $\widetilde{\mathcal{M}}_\tau$ coincide with the images of $\Phi(\,\cdot\,,\tau)$. Hence $\widetilde{\mathcal{M}}_\tau$ satisfy (9.7) by definition of Φ. $\qquad\square$

The next step is to classify the homothetically shrinking solutions, i.e., the immersions which satisfy (9.8). Such a classification is known under the additional hypothesis of mean convexity. For $n = 1$ the immersed curves with positive curvature satisfying (9.8) have been classified by Mullins in [54] and by Abresch and Langer in [1]; there is an infinite family of them, which all have self-intersections except for the circle. In higher dimension, it was shown in [42] that the unique compact solution to (9.8) with positive mean curvature is the sphere. Finally, in [43] it was proved that there are no non-compact solutions except for the products of compact ones times a flat factor. Summing up, we obtain the following conclusion.

Corollary 9.3. *Let $\mathcal{M}_t \subset \mathbb{R}^{n+1}$ be a mean curvature flow of closed mean convex hypersurfaces having a singularity of type I as $t \to T$, and let $\widetilde{\mathcal{M}}_\tau$ be a limit of rescalings as in the previous section. Then $\widetilde{\mathcal{M}}_\tau$ is a family of homothetically shrinking surfaces. More precisely, the surfaces $\widetilde{\mathcal{M}}_\tau$ are either of the form $S_\tau^{n-k} \times \mathbb{R}^k$ for some $0 \le k \le n-1$, where S_τ^{n-k} is an $(n-k)$-dimensional shrinking sphere, or of the form $G_\tau \times \mathbb{R}^{n-1}$, where G_τ is a homothetically shrinking curve in the plane belonging to the family classified in [1, 54].*

We point out that the above result is independent of the convexity estimates of Theorem 7.1. In particular, for type I singularities Corollary 8.4 can be obtained without using such estimates. For type II singularities, instead, the monotonicity formula does not yield useful information, and the convexity estimates are essential. The result which can be obtained in this case is the following.

Corollary 9.4. *Let $\mathcal{M}_t, \widetilde{\mathcal{M}}_\tau$ be as in the previous statement, in the case of a type II singularity. Then $\widetilde{\mathcal{M}}_\tau = \Gamma_\tau^{n-k} \times \mathbb{R}^k$, for some $0 \le k \le n-1$, where Γ_τ^{n-k} is an $(n-k)$-dimensional strictly convex translating solution to the flow.*

Proof. In addition to the convexity estimates, the proof uses a result by Hamilton [35] which says that any strictly convex eternal solution of the mean curvature flow is necessarily a translating solution. Applying this result to the surfaces Γ_τ^k of Theorem 8.5 we obtain our statement. $\qquad\square$

10 Cylindrical and gradient estimates

From now on, we consider the mean curvature flow of hypersurfaces which have dimension $n \ge 3$ and are uniformly two-convex, that is, satisfy $\lambda_1 + \lambda_2 \ge \alpha H$

everywhere for some $\alpha > 0$. As we have seen in Proposition 5.8, such a property is preserved by the flow. This is the class of surfaces for which a flow with surgeries has been constructed in [48]. In this section we present two estimates which are fundamental in the surgery procedure. The first one is the following.

Theorem 10.1. *Let $\mathcal{M}_t$, with $t \in [0, T)$, be a family of closed two-convex surfaces evolving by mean curvature flow. Then, for any $\eta > 0$, there exists a constant C_η such that*

$$|\lambda_1| \le \eta H \implies |\lambda_j - \lambda_k| \le c\eta H + C_\eta, \qquad j, k > 1$$

everywhere on $\mathcal{M}_t$, for $t \in [0, T)$, where c only depends on n.

We call the above result a *cylindrical estimate* because it shows that, at a point where H is large and λ_1/H is small, the Weingarten operator is close to the one of a cylinder, since all eigenvalues are close to each other except for λ_1 which is small. Such a property is important because in the surgery procedure one needs to detect regions of the surface which are close to a cylinder.

To obtain this estimate, we consider again the quotient $|A|^2/H^2$ which was used in the proof of Theorem 6.1. Observe that on a cylinder $\mathbb{R} \times S^{n-1}$ we have $|A|^2/H^2 \equiv 1/(n-1)$. A kind of converse implication also holds, namely: if at one point we have $|A|^2/H^2 = 1/(n-1)$ and in addition $\lambda_1 = 0$, then necessarily $\lambda_2 = \cdots = \lambda_n$. In fact, we have the identity

$$|A|^2 - \frac{1}{n-1}H^2 = \frac{1}{n-1}\left(\sum_{1 < i < j} (\lambda_i - \lambda_j)^2 + \lambda_1(n\lambda_1 - 2H) \right). \qquad (10.1)$$

In view of this equality, the estimate of Theorem 10.1 is an immediate consequence of the next result [48, Theorem 5.3].

Theorem 10.2. *Let $\mathcal{M}_t$, $t \in [0, T)$, be a closed two-convex solution of mean curvature flow. Then, for any $\eta > 0$, there exists a constant $C_\eta > 0$ such that*

$$|A|^2 - \frac{H^2}{n-1} \le \eta H^2 + C_\eta$$

on $\mathcal{M}_t$ for any $t \in [0, T)$.

Sketch of the proof. Let us consider, for $\eta \in \mathbb{R}$ and $\sigma \in [0, 2]$, the function

$$f_{\sigma,\eta} = \frac{|A|^2 - (\frac{1}{n-1} + \eta)H^2}{H^{2-\sigma}}. \qquad (10.2)$$

Such a function is very similar to the f_σ considered in the proof of Theorem 6.1, and it satisfies the same inequality (6.1). However, in this case we do not have a bound from below for Z analogous to (6.6). In fact, Z can be negative on non-convex surfaces. A typical example is when $\lambda_1 < 0$ and $\lambda_2 = \cdots = \lambda_n > 0$; then $Z < 0$, even if $|\lambda_1|$ is small compared to the other curvatures.

However, using also the convexity estimate of Theorem 7.2, we can show [48, Lemma 5.2] that there exists a constant $\gamma_1 > 0$ with the following property: for any $\delta > 0$ there exists K_δ such that

$$Z \geq \gamma_1 H^2 \left(|A|^2 - \frac{1}{n-1} H^2 - \delta H^2 \right) - K_\delta H^3 \tag{10.3}$$

on $\mathcal{M}_t$ for any $t > 0$. As in the proof of Theorem 7.1, we aim at estimating the L^p norms of the positive part of $f_{\sigma,\eta}$. Therefore, we only need to consider the points where the $f_{\sigma,\eta} > 0$, i.e., such that $|A|^2 - \frac{H^2}{n-1} \geq \eta H^2$. Thus, if we choose $\delta = \eta/2$ in (10.3) the first term is positive and the only negative contribution to the right-hand side is the last term, which has lower order. It turns out that this estimate can play the same role as (6.6), and that it is possible to obtain an upper bound for $f_{\sigma,\eta}$ by a procedure similar to the one of Theorem 6.2. The bound on $f_{\sigma,\eta}$ easily implies the estimate of Theorem 10.2. $\qquad\square$

We next describe an estimate for the gradient of the curvature for our evolving surfaces. Compared to the gradient estimates for mean curvature flow already available in the literature, e.g., [18, 22], this one is peculiar because it is a pointwise estimate, which does not depend on the maximum of the curvature in a suitable neighbourhood. A similar estimate has been obtained by Perelman [55, 56] for the Ricci flow by a completely different approach.

Theorem 10.3. *Let $\mathcal{M}_t$, $t \in [0, T)$, be a closed n-dimensional two-convex solution of mean curvature flow, with dimension $n \geq 3$. Then there is a constant $\gamma_2 = \gamma_2(n)$ and a constant $\gamma_3 = \gamma_3(n, \mathcal{M}_0)$ such that the flow satisfies the uniform estimate*

$$|\nabla A|^2 \leq \gamma_2 |A|^4 + \gamma_3 \tag{10.4}$$

for every $t \in [0, T)$.

Proof. The result is obtained by applying the maximum principle to a suitable function that we are going to introduce. An important tool in the proof is the inequality [38, Lemma 2.1], valid for any immersed hypersurface,

$$|\nabla A|^2 \geq \frac{3}{n+2} |\nabla H|^2. \tag{10.5}$$

Observe that $\frac{3}{n+2} > \frac{1}{n-1}$ if $n \geq 3$. Let us set

$$\kappa_n = \frac{1}{2} \left(\frac{3}{n+2} - \frac{1}{n-1} \right).$$

By Theorem 10.2, there exists $C_0 > 0$ such that

$$\left(\frac{1}{n-1} + \kappa_n \right) H^2 - |A|^2 + C_0 \geq 0.$$

Let us set

$$g_1 = \left(\frac{1}{n-1} + \kappa_n\right) H^2 - |A|^2 + 2C_0, \qquad g_2 = \frac{3}{n+2} H^2 - |A|^2 + 2C_0.$$

Then we have $g_2 > g_1 \geq C_0$, and so $g_i - 2C_0 = 2(g_i - C_0) - g_i \geq -g_i$ for $i = 1, 2$. Using the evolution equations for $|A|^2, H^2$ (see Lemma 5.1) and inequality (10.5), we find

$$\begin{aligned}
\frac{\partial}{\partial t} g_1 - \Delta g_1 &= -2\left(\left(\frac{1}{n-1} + \kappa_n\right)|\nabla H|^2 - |\nabla A|^2\right) + 2|A|^2 (g_1 - 2C_0) \\
&\geq 2\left(1 - \frac{n+2}{3}\left(\frac{1}{n-1} + \kappa_n\right)\right)|\nabla A|^2 - 2|A|^2 g_1 \qquad (10.6) \\
&= 2\kappa_n \frac{n+2}{3}|\nabla A|^2 - 2|A|^2 g_1.
\end{aligned}$$

Similarly,

$$\frac{\partial}{\partial t} g_2 - \Delta g_2 = -2\left(\frac{3}{n+2}|\nabla H|^2 - |\nabla A|^2\right) + 2|A|^2 (g_2 - 2C_0) \geq -2|A|^2 g_2. \quad (10.7)$$

In addition (see [38, Theorem 7.1]),

$$\frac{\partial}{\partial t}|\nabla A|^2 - \Delta|\nabla A|^2 \leq -2|\nabla^2 A|^2 + c_n|A|^2|\nabla A|^2, \qquad (10.8)$$

for a constant c_n depending only on n. Using the above relations one obtains, after a straightforward computation, the following inequality for the quotient $|\nabla A|^2/g_1 g_2$:

$$\begin{aligned}
&\frac{\partial}{\partial t}\left(\frac{|\nabla A|^2}{g_1 g_2}\right) - \Delta\left(\frac{|\nabla A|^2}{g_1 g_2}\right) - \frac{2}{g_2}\left\langle \nabla g_2, \frac{|\nabla A|^2}{g_1 g_2}\right\rangle \\
&\leq \frac{|\nabla A|^2 |A|^2}{g_1 g_2}\left((c_n + 4) - 2\kappa_n^2 \frac{n+2}{3n}\frac{|\nabla A|^2}{g_1 g_2}\right).
\end{aligned}$$

Thus, at any point where $|\nabla A|^2/g_1 g_2$ attains a new maximum, we have

$$\frac{|\nabla A|^2}{g_1 g_2} \leq \frac{3n(c_n + 4)}{2\kappa_n^2(n+2)}.$$

Using the definition of g_1, g_2, this easily yields our assertion. □

Once the estimate for $|\nabla A|$ is obtained, it is easy to obtain similar estimates for the higher order derivatives, as well as the time derivatives.

11 Mean curvature flow with surgeries

In this section we describe the mean curvature flow with surgeries which has been defined in [48] for two-convex surfaces of dimension $n \geq 3$. Such a construction is inspired by the one which was introduced by Hamilton [37] for the Ricci flow and which enabled Perelman [56] to prove the geometrization conjecture for three-dimensional manifolds.

The aim of the flow with surgeries is to define a continuation of the smooth flow past the first singular time until the surface has approached some canonical limit and we are able to determine its topological type. We remark that this purpose would be hard to obtain by means of the weak solutions of the mean curvature flow mentioned previously in these notes. In fact, weak solutions have low regularity past the singular time and it is difficult to analyze their topological behaviour. The flow with surgeries is based on a different strategy. While weak solutions are non-smooth surfaces solving the equation exactly, solutions with surgeries are smooth surfaces solving the equation up to certain errors introduced at given times. At these times, the topological type of the surface may change, but in a controlled way. Thus, we deal with a smooth surface throughout the evolution, and it is possible to keep track of the changes of topology.

More precisely, the flow with surgeries follows this approach. If at the singular time T the whole surface vanishes, then we do nothing and consider the flow terminated at time T. We assume that we have enough knowledge of the formation of singularities that we can tell the possible topological type of a surface that vanishes completely at the singular time. If the surface instead does not vanish at time T, we stop the flow at some time T_0 slightly smaller than T. We remove from the surface $\mathcal{M}_{T_0}$ one or more regions with large curvature and replace them with more regular ones. Such an operation is called a *surgery*. It may possibly disconnect the surface into different components. The flow is then restarted for each component until a new singular time is approached. The procedure is repeated until each component vanishes.

In order to define rigorously such a procedure, one needs to specify the geometric properties of the regions that are removed in the surgeries and of the ones that are added as a replacement. To this purpose, one introduces the notion of *neck*. The precise definition is given in [48]; roughly speaking, a neck is a portion of a surface which is close, up to a homothety and a rigid motion, to a standard cylinder $[a, b] \times S^{n-1}$. The surgeries which we consider consist of removing a neck and of replacing it with two regions diffeomorphic to disks which fill smoothly the two holes left at the two ends of the removed neck. In this way we can describe precisely the possible changes of topology of the surface. In fact, the surgery is the inverse of the operation which is called a *direct sum* in topology. If we are able to show that, after a finite number of surgeries, all remaining components have a known topology, then the initial surface is necessarily diffeomorphic to the direct sum of components with those properties. It turns out that this program can be carried out, and that the following result can be obtained.

Theorem 11.1. *Let $\mathcal{M}_0 \subset \mathbb{R}^{n+1}$ be a closed immersed n-dimensional two-convex hypersurface, with $n \geq 3$. Then there is a mean curvature flow with surgeries with initial value $\mathcal{M}_0$ such that, after a finite number of surgeries, the remaining components are diffeomorphic either to S^n or to $S^{n-1} \times S^1$.*

Due to the structure of our surgeries, the theorem implies that the initial manifold is the connected sum of finitely many components diffeomorphic to S^n or to $S^{n-1} \times S^1$. Recalling that the connected sum with S^n leaves the topology unchanged, we obtain the following classification of two-convex hypersurfaces.

Corollary 11.2. *Any smooth closed n-dimensional two-convex immersed surface $\mathcal{M} \subset \mathbb{R}^{n+1}$ with $n \geq 3$ is diffeomorphic either to S^n or to a finite connected sum of $S^{n-1} \times S^1$.*

Topological results on k-convex surfaces were already known in the literature (see, e.g., [69]). However, these results were based on Morse theory and only ensured homotopy equivalence. Another consequence of our construction is the following Schoenflies type theorem for simply connected two-convex surfaces.

Corollary 11.3. *Any smooth closed simply connected n-dimensional two-convex embedded surface $\mathcal{M} \subset \mathbb{R}^{n+1}$ with $n \geq 3$ is diffeomorphic to S^n and bounds a region whose closure is diffeomorphic to a smoothly embedded $(n+1)$-dimensional standard closed ball.*

The proof of Theorem 11.1 is quite long and technical. Let us only explain, at an intuitive level, how the approach works and which is the role of the two-convexity.

A compact two-convex surface is also uniformly two-convex, i.e., it satisfies $\lambda_1 + \lambda_2 \geq \alpha H$ everywhere for some $\alpha > 0$. As we have seen in Proposition 5.8, this property is preserved by the flow. It is also scale invariant, and therefore any smooth limit of rescalings must satisfy the same inequality. This restricts the number of possibilities in Corollaries 9.3, 9.4, since the only uniformly two-convex limits are the sphere S_t^n, the cylinder $S_t^{n-1} \times \mathbb{R}$ and the n-dimensional translating solutions Γ_t^n.

If the limit is a sphere, then we do not need to perform any surgery on the surface (or on that component of the surface) since we know that it is a convex component shrinking to a round point. If the limit is a cylinder, then we have the right geometric structure to perform a surgery. The case of a translating solution Γ_t^n, which corresponds to type II singularities, is less obvious. The typical profile of a translating solution is in fact paraboloidal, rather than cylindrical. However, far away from the vertex a paraboloid looks more and more similar to a cylinder. Thus, in this case we perform the surgery not at the point where the curvature is the largest, but in a region nearby, where the curvature is still quite large and the shape of the surface is closer to a cylinder.

The precise implementation of these ideas is quite long and technical. The estimates of the previous sections play a fundamental role to prove the existence of necks which are suitable for the surgery procedure. It is also essential that the surgeries do not alter the validity of the estimates, so that they hold with the same constants even after the modifications at the surgery times. This allows us to define a flow with surgeries where the curvature remains uniformly bounded. Such a flow necessarily terminates after a finite number of steps, because the area decreases by a fixed amount with each surgery, and is decreasing along the smooth flow by (4.3).

Acknowledgments

The author wishes to thank V. Miquel and J. Porti for the opportunity of contributing to the Advanced Course at the CRM in March 2008. He also wishes to thank the director J. Bruna and the staff people at the CRM for their hospitality.

Bibliography

[1] U. Abresch, J. Langer, *The normalized curve shortening flow and homothetic solutions*, J. Differential Geom. **23** (1986), 175–196.

[2] R. Alessandroni, C. Sinestrari, *Evolution of hypersurfaces by powers of the scalar curvature* (2009), preprint.

[3] S. J. Altschuler, S. B. Angenent, Y. Giga, *Mean curvature flow through singularities for surfaces of rotation*, J. Geom. Analysis **5** (1995), 293–358.

[4] L. Ambrosio, *Geometric evolution problems, distance function and viscosity solutions*, in: G. Buttazzo, A. Marino, M. K. V. Murthy (eds.), Calculus of Variations and Partial Differential Equations (Pisa, 1996), Springer, Berlin (2000).

[5] L. Ambrosio, H. M. Soner, *Level set approach to mean curvature flow in any codimension*, J. Differential Geom. **43** (1996), 693–737.

[6] B. Andrews, *Contraction of convex hypersurfaces in Euclidean space*, Calc. Variations **2** (1994), 151–171.

[7] B. Andrews, *Gauss curvature flow: the fate of the rolling stones*, Invent. Math. **138** (1999), 151–161.

[8] B. Andrews, *Motion of hypersurfaces by Gauss curvature*, Pacific J. Math. **195** (2000), 1–36 .

[9] S. B. Angenent, *On the formation of singularities in the curve shortening flow*, J. Differential Geom. **33** (1991), 601–633.

[10] S. B. Angenent, *Shrinking doughnuts* in: Nonlinear Diffusion Equations and Their Equilibrium States (Gregynog, 1989), Birkhäuser, Boston (1992).

[11] S. B. Angenent, J. J. L. Velázquez, *Degenerate neckpinches in mean curvature flow*, J. Reine Angew. Math. **482** (1997), 15–66.

[12] K. A. Brakke, The Motion of a Surface by Its Mean Curvature, Princeton University Press, Princeton (1978).

[13] E. Cabezas-Rivas, V. Miquel, *Volume preserving mean curvature flow in the hyperbolic space*, Indiana Univ. Math. J. **56** (2007), 2061–2086.

[14] Y. G. Chen, Y. Giga, S. Goto, *Uniqueness and existence of viscosity solutions of generalized mean curvature flow equation*, J. Differential Geom. **33** (1991), 749–786.

[15] B. Chow, *Deforming convex hypersurfaces by the nth root of the Gaussian curvature*, J. Differential Geom. **22** (1985), 117–138.

[16] B. Chow, *Deforming convex hypersurfaces by the square root of the scalar curvature*, Invent. Math. **87** (1987), 63–82.

[17] J. Clutterbuck, O. C. Schnürer, F. Schulze, *Stability of translating solutions to mean curvature flow*, Calc. Var. Partial Differential Equations **29** (2007), 281–293.

[18] T. Colding, W. P. Minicozzi, *Sharp estimates for mean curvature flow of graphs*, J. Reine Angew. Math. **574** (2004), 187–195.

[19] P. Daskalopoulos, R. S. Hamilton, *The free boundary in the Gauss curvature flow with flat sides*, J. Reine Angew. Math. **510** (1999), 187–227.

[20] K. Ecker, *A local monotonicity formula for mean curvature flow*, Ann. of Math. **154** (2001), 503–525.

[21] K. Ecker, Regularity Theory for Mean Curvature Flow, Birkhäuser, Boston (2004).

[22] K. Ecker, G. Huisken, *Interior estimates for hypersurfaces moving by mean curvature*, Invent. Math. **105** (1991), 547–569.

[23] L. C. Evans, J. Spruck, *Motion of level sets by mean curvature, I*, J. Differential Geom. **33** (1991) 635–681.

[24] L. C. Evans, J. Spruck, *Motion of level sets by mean curvature, II*, Trans. Amer. Math. Soc. **330** (1992) 321–332.

[25] W. J. Firey, *Shapes of worn stones*, Mathematica **21** (1974), 1–11.

[26] M. Gage, R. S. Hamilton, *The heat equation shrinking convex plane curves*, J. Differential Geom. **23** (1986), 69–96.

[27] C. Gerhardt, Curvature Problems, International Press, Sommerville, MA (2006).

[28] Y. Giga, Surface Evolution Equations: A Level Set Approach, Birkhäuser, Basel (2006).

[29] Y. Giga, S. Goto, *Geometric evolution of phase boundaries*, in: M. E. Gurtin and G. B. McFadden (eds.), On the Evolution of Phase Boundaries, IMA Volumes in Mathematics and Applications 43, Springer-Verlag, New York (1992).

[30] M. A. Grayson, *The heat equation shrinks embedded plane curves to round points*, J. Differential Geom. **26** (1987), 285–314.

[31] M. A. Grayson, *Shortening embedded curves*, Ann. of Math. **129** (1989), 71–111.

[32] M. A. Grayson, *A short note on the evolution of a surface by its mean curvature*, Duke Math. J. **58** (1989), 555–558.

[33] R. S. Hamilton, *Three-manifolds with positive Ricci curvature*, J. Differential Geom. **17** (1982), 255–306.

[34] R. S. Hamilton, *Four-manifolds with positive curvature operator*, J. Differential Geom. **24** (1986), 153–179.

[35] R. S. Hamilton, *The Harnack estimate for the mean curvature flow*, J. Differential Geom. **41** (1995), 215–226.

[36] R. S. Hamilton, *Formation of singularities in the Ricci flow*, Surveys in Differential Geometry **2** (1995), 7–136, International Press, Boston.

[37] R. S. Hamilton, *Four-manifolds with positive isotropic curvature*, Comm. Anal. Geom. **5** (1997), 1–92.

[38] G. Huisken, *Flow by mean curvature of convex surfaces into spheres*, J. Differential Geom. **20** (1984), 237–266.

[39] G. Huisken, *Contracting convex hypersurfaces in Riemannian manifolds by their mean curvature*, Invent. Math. **84** (1986), 463–480.

[40] G. Huisken, *Deforming hypersurfaces of the sphere by their mean curvature*, Math. Z. **195** (1987), 205–219.

[41] G. Huisken, *The volume preserving mean curvature flow*, J. Reine Angew. Math. **382** (1987) 35–48.

[42] G. Huisken, *Asymptotic behaviour for singularities of the mean curvature flow*, J. Differential Geom. **31** (1990), 285–299.

[43] G. Huisken, *Local and global behaviour of hypersurfaces moving by mean curvature*, Proceedings of Symposia in Pure Mathematics **54** (1993), 175–191.

[44] G. Huisken, T. Ilmanen, *The inverse mean curvature flow and the Riemannian Penrose inequality*, J. Differential Geom. **59** (2001), 353–437.

[45] G. Huisken, A. Polden, *Geometric evolution equations for hypersurfaces*, in: S. Hidebrandt, M. Struwe (eds.), Calculus of Variations and Geometric Evolution Problems (Cetraro, 1996), Springer-Verlag, Berlin, Heidelberg (1999).

[46] G. Huisken, C. Sinestrari, *Mean curvature flow singularities for mean convex surfaces*, Calc. Variations **8** (1999), 1–14.

[47] G. Huisken, C. Sinestrari, *Convexity estimates for mean curvature flow and singularities of mean convex surfaces*, Acta Math. **183** (1999), 45–70.

[48] G. Huisken, C. Sinestrari, *Mean curvature flow with surgeries of two-convex hypersurfaces*, Invent. Math. **175** (2009), 137–221.

[49] T. Ilmanen, *Elliptic regularization and partial regularity for motion by mean curvature*, Mem. Amer. Math. Soc. **108** (1994), no. 520.

[50] A. Lunardi, Analytic semigroups and optimal regularity in parabolic problems, Birkhäuser, Basel (1995).

[51] M. Marcus, L. Lopes, *Inequalities for symmetric functions and hermitian matrices*, Canad. J. Math. **9** (1957), 305–312.

[52] J. A. McCoy, *Mixed volume preserving curvature flows*, Calc. Variations **24** (2005), 131–154.

[53] J. H. Michael, L. M. Simon, *Sobolev and mean value inequalities on generalized submanifolds of* $\mathrm{I\!R}^n$, Comm. Pure Appl. Math. **26** (1973), 361–379.

[54] W. W. Mullins *Two-dimensional motion of idealised grain boundaries*, J. Appl. Phys. **27** (1956), 900–904.

[55] G. Perelman, *The entropy formula for the Ricci flow and its geometric applications*, preprint (2002).

[56] G. Perelman, *Ricci flow with surgery on three-manifolds*, preprint (2003).

[57] O. C. Schnürer, *Surfaces contracting with speed* $|A|^2$, J. Differential Geom. **71** (2005), 347–363.

[58] O. Schnürer, *Geometric evolution equations*, Lecture notes of the Alpbach Summer School (2007), available on the web page of the author.

[59] F. Schulze, *Convexity estimates for flows by powers of the mean curvature (with an appendix by F. Schulze and O. Schnürer)*, Ann. Sc. Norm. Super. Pisa Cl. Sci. (5) **5** (2006), 261–277.

[60] L. Simon, Lectures on Geometric Measure Theory, Proceedings of the CMA, vol. 3. Australian National University, Canberra (1983).

[61] K. Smoczyk *Self-shrinkers of the mean curvature flow in arbitrary codimension*, Int. Math. Research Notices **48** (2005), 2983–3004.

[62] P. E. Souganidis, *Front propagation: theory and applications*, in: I. Capuzzo Dolcetta and P. L. Lions (eds.), Viscosity Solutions and Applications, Springer-Verlag, Berlin (1997), 186–242.

[63] K. Tso, *Deforming a hypersurface by its Gauss-Kronecker curvature*, Comm. Pure Appl. Math. **38** (1985), 867–882.

[64] J. Urbas, *An expansion of convex hypersurfaces*, J. Differential Geom. **33** (1991), 91–125.

[65] M. T. Wang, *Mean curvature flow in higher codimension*, Proceedings of the Second International Congress of Chinese Mathematicians (2002), available at www.arxiv.org.

[66] M. T. Wang, *Some recent developments in Lagrangian mean curvature flows*, in: H. D. Cao and S. T. Yau (eds.), Surveys in Differential Geometry, vol. XII, Geometric Flow, International Press, Sommerville (2008).

[67] X. J. Wang, *Convex solution to the mean curvature flow*, preprint (2004), available at www.arXiv.org.

[68] B. White *The nature of singularities in mean curvature flow of mean-convex sets*, J. Amer. Math. Soc. **16** (2002), 123–138.

[69] H. Wu, *Manifolds of partially positive curvature*, Indiana Univ. Math. J. **36** (1987), 525–548.

Part II

Geometric Flows, Isoperimetric Inequalities and Hyperbolic Geometry

Manuel Ritoré

To Carmen and Álvaro

Preface

Given a Riemannian manifold M and some volume v less than the volume of M, one can look for the regions in M, if they exist, of volume v and minimum perimeter. Classically, this problem has been treated by using *isoperimetric inequalities*, which are nothing but relations between the volume of a set and its boundary area. A fundamental example is the classical isoperimetric inequality in the three-dimensional Euclidean space

$$A^3 \geqslant 36\pi\, V^2,$$

where A, V denote the boundary area and the volume of a given set. This minimization problem can be considered not only on Riemannian manifolds, but also in spaces where suitable notions of volume and perimeter can be defined, such as metric measure spaces, where the volume is given by the measure, and the boundary area is its Minkowski content, defined in terms of the measure and the distance.

Isoperimetric inequalities have been proved by many different methods: complex analysis, symmetrizations, Sobolev inequalities, and classical calculus of variations, amongst many others. A natural way of proving these inequalities is to deform the domain while keeping controlled the ratio between the perimeter and some function of the volume. The classical approach by inner equidistant curves is an example of this procedure. Geometric flows, obtained from solutions of geometric parabolic equations, can be considered an alternative tool to prove these isoperimetric inequalities. On the other hand, isoperimetric inequalities can help in treating several aspects of convergence of these flows.

Isoperimetric inequalities are a valuable tool in Riemannian geometry, in particular in hyperbolic geometry. The use of variational problems to study geometric and topological properties of manifolds is a very interesting field. Works by Lawson, Meeks, Meeks and Yau, Meeks, Simon and Yau, Pitts and Rubinstein, on minimal surfaces pointed in this direction.

Our goals, in writing these notes, are twofold. First we shall present recent developments in the field of isoperimetric inequalities, mostly based on the use of geometric flows. These techniques not only shed light on the proof of classical results, but provide the machinery to prove new isoperimetric inequalities. Second, we shall give applications of isoperimetric inequalities to hyperbolic geometry.

The structure of these notes is as follows:

Chapter 1. Here we give a proof of the classical isoperimetric inequality in Euclidean space of any dimension. Our goal is to keep these notes self-contained, and to provide some basic tools and notation for further reference. We present two proofs of the classical isoperimetric inequality. The first, which dates back to Schwarz, makes use of a symmetrization technique to reduce the problem to the comparison of sets of revolution around a given line. Then a new calibration argument is given to prove that the spheres have less perimeter than the rotationally symmetric candidates. A second proof by Montiel and Ros, based on Gromov's ideas and Heintze–Karcher techniques, is presented.

Chapter 2. We use Grayson's result on the convergence of the curve shortening flow on Riemannian surfaces to prove 2-dimensional isoperimetric inequalities. The theory of curve shortening flow, under certain geometric restrictions on the surface, is complete enough to replace the classical method by parallels, as was shown by Benjamini and Cao. We give their results on planes, simplifying part of their arguments by using the avoidance principle for the curve shortening flow. Further results by Morgan and Johnson on spheres are given. Recent developments stemming from Benjamini and Cao's work are briefly discussed.

Chapter 3. We review the few known results on applications of the mean curvature flow to obtain isoperimetric inequalities in higher dimensions. We focus especially on the results by Schulze on weak H^k flows. A key tool in Schulze's method is obtaining lower bounds for the Willmore functional

$$\int_\Sigma H^n \, d\Sigma$$

for hypersurfaces $\Sigma \subset \mathbb{R}^{n+1}$. We discuss how to prove isoperimetric inequalities directly from lower bounds of the Willmore functional, an argument that goes back to Almgren. These techniques can also be applied to the exterior of a convex set.

Chapter 4. In this chapter we give two applications of the theory of isoperimetric inequalities to hyperbolic geometry. First we present a result by Bachman, Cooper and White, giving a bound on the injectivity radius of a compact, connected, orientable, hyperbolic 3-manifold in terms of the Heegaard genus of the manifold. We also review the result by Adams and Morgan on the isoperimetric profile for small volumes of an $(n + 1)$-dimensional cusped hyperbolic manifold.

Many topics are not covered by these notes, or referred to only briefly. They include:

1. The theory of constant mean curvature hypersurfaces and the notion of stable constant mean curvature hypersurfaces. These are the second-order minima of area under a (second-order) volume constraint. It is generally believed that the theory of stable hypersurfaces could be the most fruitful approach towards the resolution of the isoperimetric problem in manifolds with a certain topological complexity. Moreover, the local minimization property of these hypersurfaces is perhaps a more realistic hypothesis than the global one satisfied by solutions of the isoperimetric problem.

2. Analytic isoperimetric inequalities, Sobolev inequalities and Faber–Krahn type inequalities concerning bounds for eigenvalues of the Laplacian.

3. Isoperimetric problems with additional constraints, such as fixed boundary, free boundary, regions confined inside a given sets, or perimeter-minimizing enclosures of several regions, amongst others.

4. Geometric properties of isoperimetric regions such as connectedness, convexity, or bounds on the genus of the boundary in 3-dimensional manifolds.

5. Extensions of isoperimetric inequalities to non-Riemannian spaces, such as sub-Riemannian manifolds, Carnot groups, manifolds with density, metric measure spaces in general, graphs or singular spaces.

Remark on notation, conventions, and bibliography

Our notation and terminology are for the most part standard. The Euclidean space of dimension n and its unit sphere will be denoted by $\mathbb{R}^n$ and $\mathbb{S}^{n-1}$, respectively. For a given set Ω, $V(\Omega)$ will usually denote its volume and $A(\partial\Omega)$ its boundary area. The n-dimensional area of $\mathbb{S}^n$ will be indicated by c_n. The unit ball centered at the origin in $\mathbb{R}^n$ will be denoted by $\mathbb{B}^n$ and its n-dimensional volume by ω_n. The k-dimensional Lebesgue measure is $\mathcal{L}^k$. A Riemannian manifold (M, g) will be usually mentioned as M, without reference to the Riemannian metric. Manifolds will always be of class C^∞ with C^∞ Riemannian metrics.

The list of references is not intended to be exhaustive. We refer the reader to the more complete books by Chavel [24], [25], and the treatise by Burago and Zalgaller [19]. The recent fourth edition of Frank Morgan's *Geometric Measure Theory: A Beginner's Guide* [68] has been used when writing this work.

Acknowledgements

These notes are an elaborated version of an advanced course given by the author at the Centre de Recerca Matemàtica in March 2008. The course was included in a research semester on Geometric Flows and Hyperbolic Geometry. The author is grateful to the organizers, Vicente Miquel and Joan Porti, for their support and encouragement. He would also like to thank Manuel Castellet for offering the possibility of publishing these notes.

These notes are based on the work of a large number of authors and have benefitted from feedback and comments from many colleagues. In particular the author is grateful to César Rosales and Frank Morgan, for their careful reading of these notes and making many valuable suggestions. To Gary Lawlor, for sending his notes on metacalibrations. To Stratos Vernadakis, for pointing out a few precise comments, and Antonio Cañete and Matteo Galli, and to the members of the Department of Geometry and Topology of the University of Granada, in particular to Antonio Ros, Sebastián Montiel, Francisco Urbano and Joaquín Pérez. Financial support has been provided by the Centre de Recerca Matemàtica and the MEC-Feder grant MTM2007-61919.

Chapter 1

The classical isoperimetric inequality in Euclidean space

1.1 Introduction

The isoperimetric inequality in $(n+1)$-dimensional Euclidean space reads

$$A^{n+1} \geqslant \frac{A(\mathbb{S}^n)^{n+1}}{V(\mathbb{B}^{n+1})^n} V^n, \tag{1.1}$$

where V is the volume of an open set $\Omega \subset \mathbb{R}^{n+1}$ with smooth boundary and A its boundary area. Equality holds in (1.1) if and only if Ω is a Euclidean ball. Inequality (1.1) shows that, amongst all regions with the same boundary area, the Euclidean balls have maximum volume. This minimization problem is usually referred to as an *isoperimetric problem*, and can be stated in quite general spaces, where appropriate notions of area and volume can be defined. Inequality (1.1) also shows that, amongst all sets with the same volume, Euclidean balls have minimum boundary area. In the next section we shall introduce a minimal set of definitions and results needed to treat these variational problems on Riemannian manifolds.

1.2 Preliminaries

1.2.1 Area and volume

On a Riemannian manifold $(M, \langle \, , \, \rangle)$, one may define a Riemannian measure [25, p. 158], which coincides with the Lebesgue measure in Euclidean spaces. The *volume* $V(\Omega)$ of $\Omega \subset M$ will be defined as the Riemannian measure of Ω. The volume element will be denoted by dM, so that the integral of the function f with respect to this measure will be $\int_M f \, dM$. If $\Sigma \subset M$ is a submanifold of dimension

$k \leqslant \dim M - 1$, the *area* $A(\Sigma)$ of Σ will be the Riemannian measure associated to the Riemannian metric induced on Σ by M. The associated Riemannian volume element will be denoted by $d\Sigma$. If Σ is 1-dimensional, we shall speak of *length* instead of area and we will denote it by L. Moreover, if M is a surface, then the term volume will be replaced by *area*.

Some other notions of volume and area can be defined on M. Let $k \in \mathbb{N}$ and $\delta > 0$. For $A \subset M$, define

$$\mathcal{H}_\delta^k(A) = \inf\left\{ \sum_{j=1}^\infty \alpha_k \left(\frac{\mathrm{diam}(C_j)}{2} \right)^k : A \subset \bigcup_{j=1}^\infty C_j, \ \mathrm{diam}(C_j) \leqslant \delta \right\},$$

where α_k is the volume of the unit ball in $\mathbb{R}^k$. Then the *k-dimensional Hausdorff measure* of A is defined by

$$\mathcal{H}^k(A) = \lim_{\delta \to 0} \mathcal{H}_\delta^k(A) = \sup_{\delta > 0} \mathcal{H}_\delta^k(A).$$

The area of a k-dimensional submanifold of M coincides with its k-dimensional Hausdorff measure; see [19, 13.2], [34, Ch. 2]. Observe that the Hausdorff measures can be considered on any metric space.

Another notion of boundary area for sets is the Minkowski content. Let $\Omega \subset M$ be any set. For $r > 0$, consider the closed tubular neighborhood of radius r of Ω given by

$$\Omega_r = \{p \in M : d(p, \Omega) \leqslant r\},$$

where d is the Riemannian distance on M. The *Minkowski content* of Ω is defined by

$$\mathcal{M}(\Omega) = \liminf_{r \downarrow 0} \frac{V(\Omega_r) - V(\Omega)}{r}.$$

For the definition of Minkowski content one needs a measure and a distance. Hence this notion of boundary area can be introduced in metric measure spaces; see [19, Ch. 2].

Finally, consider the set $\mathfrak{X}_0(M)$ of smooth vector fields on M with compact support. Given $\Omega \subset M$, the *perimeter* of Ω, [42], is defined by

$$\mathcal{P}(\Omega) = \sup\left\{ \int_\Omega \mathrm{div}\, X \, dM : X \in \mathfrak{X}_0(M), \ |X| \leqslant 1 \right\},$$

where $|X|$ is the supremum norm $\sup\{|X|_p : p \in M\}$. An equivalent definition, sometimes called the *geometric perimeter*, is defined by

$$\inf\{\liminf A(\partial \Omega_i)\},$$

where $\{\Omega_i\}_{i \in \mathbb{N}}$ is any sequence of sets with smooth boundary such that the characteristic functions of Ω_i converge in L^1 to Ω. There are extensions of the notion of perimeter to sub-Riemannian manifolds, [36], [41]. The perimeter functional, unlike Hausdorff measures and Minkowski content, is lower semicontinuous with respect to L^1_{loc} convergence of the characteristic function of sets.

If Σ' is a bounded portion of a smooth hypersurface Σ enclosing a region in M, then the Minkowski content of Σ', its n-dimensional Hausdorff measure, its perimeter and its area coincide. A detailed discussion of the relation between Hausdorff measure and perimeter, and Minkowski content and perimeter, can be found in Chavel's book on isoperimetric problems [24, Ch. III, Ch. IV].

The usual results for non-negative measures, as well as the Riemannian Divergence Theorem, and the area and coarea formulas, are valid on M, and will be used throughout these notes. The reader is referred to Burago–Zalgaller [19] or Chavel [25] for further reference.

1.2.2 Variational formulas

For the most part, this subsection is taken from Simon's book [110]. Let $\Sigma \subset M$ be a submanifold in a Riemannian manifold. Consider a vector field X on M with compact support, and let $\{\varphi_t\}_{t\in\mathbb{R}}$ be the associated one-parameter group of diffeomorphisms. Then one has [110, §9]

$$\frac{d}{dt}\bigg|_{t=0} A(\varphi_t(\Sigma)) = \int_\Sigma \text{div}_\Sigma X \, d\Sigma, \tag{1.1}$$

where the divergence of X in Σ, $\text{div}_\Sigma X$, is defined as follows: let $p \in \Sigma$, and $\{e_1, \ldots, e_k\}$ be an orthonormal basis of $T_p\Sigma$; denoting by ∇ the Levi-Civita connection on M, then

$$(\text{div}_\Sigma X)(p) = \sum_{i=1}^k \langle \nabla_{e_i} X, e_i \rangle.$$

We remark that formula (1.1) is also valid for submanifolds $\Sigma \subset M$ with non-empty boundary.

In case $\Sigma = \Omega$ is an open set in M, div_Σ is the usual divergence on M and will be denoted simply by div. In this case we have from (1.1) and the Divergence Theorem

$$\frac{d}{dt}\bigg|_{t=0} V(\varphi_t(\Omega)) = \int_\Omega \text{div}\, X \, dM = -\int_{\partial\Omega} \langle X, N \rangle \, d(\partial\Omega), \tag{1.2}$$

where N is the inner unit normal to $\partial\Omega$.

In case Σ is a hypersurface with unit vector field N normal to Σ, we have

$$\text{div}_\Sigma X = \text{div}_\Sigma X^\top + \text{div}_\Sigma X^\perp = \text{div}_\Sigma X^\top - nH\langle X, N \rangle, \tag{1.3}$$

where $X^\top$, $X^\perp$ are the tangent and normal projections of X to Σ, and

$$H(p) = -\frac{1}{n}\sum_{i=1}^n \langle \nabla_{e_i} N, e_i \rangle$$

is the mean curvature of Σ. If the vectors e_i are principal directions, then

$$-\nabla_{e_i} N = \kappa_i e_i,$$

where κ_i is the principal curvature associated to e_i, and so

$$H(p) = \frac{1}{n} \sum_{i=1}^{n} \kappa_i.$$

Assuming that the support of X is contained in the interior of Σ, using (1.3) and the Riemannian Divergence Theorem we get

$$\frac{d}{dt}\bigg|_{t=0} A(\varphi_t(\Sigma)) = -\int_\Sigma nH \langle X, N \rangle \, d\Sigma. \tag{1.4}$$

From formulas (1.2) and (1.4) we obtain the usual characterization of minimal and constant mean curvature hypersurfaces as the critical points of area, without or with a volume constraint, respectively. From these variational formulas it is also straightforward to obtain that a hypersurface $\Sigma \subset M$ enclosing a region Ω has constant mean curvature H_0 if and only if it is a critical point of the functional $A - nH_0V$ for any variation.

1.2.3   The isoperimetric profile

Let M be a smooth complete Riemannian manifold. The *isoperimetric profile* of M is the function $I_M \colon (0, V(M)) \to \mathbb{R}$ defined by

$$I_M(v) = \inf\{A(\partial\Omega) : \Omega \subset\subset M \text{ has smooth boundary}, \ V(\Omega) = v\}. \tag{1.5}$$

The isoperimetric profile gives an isoperimetric inequality in M, since any region $\Omega \subset\subset M$ with smooth boundary satisfies

$$A(\partial\Omega) \geqslant I_M(V(\Omega)).$$

This isoperimetric inequality is *optimal* in the sense that, if some function I exists so that $A(\partial\Omega) \geqslant I(V(\Omega))$ for any region $\Omega \subset\subset M$ with smooth boundary, one trivially has $I_M \geqslant I$.

Given two Riemannian manifolds M_1 and M_2 with infinite volume, their isoperimetric profiles can be compared. Inequality $I_{M_1} \geqslant I_{M_2}$ implies that the function I_{M_2} provides an isoperimetric inequality in M_1, usually non-optimal. In case that M_1 and M_2 are compact, one can compare their *normalized profiles* $J_{M_i}(t) = I_{M_i}(V(M_i)\,t)$, $i = 1, 2$, $t \in (0,1)$.

Analytic properties of the isoperimetric profile can be found in Bavard–Pansu [11], Hsiang [55], Morgan–Johnson [71], and Ros [98]. Some of them will be used later in this work.

1.2.4 Isoperimetric regions

Since any finite perimeter set can be approximated by a sequence of bounded sets with C^∞ boundary, the value $I_M(v)$ can also be computed as

$$\inf\{\mathcal{P}(\Omega) \ : \ \Omega \text{ has finite perimeter}, \ V(\Omega) = v\}.$$

Observe that a *minimizing sequence* $\{\Omega_i\}_{i\in\mathbb{N}}$ of sets with smooth boundary and volume v so that $A(\partial\Omega_i) \to I_M(v)$ may converge, in a weak topology, to some set with non-smooth boundary. This motivates the following definition: an *isoperimetric region* of volume v in M is a finite perimeter set Ω_0 so that $V(\Omega_0) = v$ and $\mathcal{P}(\Omega_0) = I_M(v)$. Existence of isoperimetric regions is not guaranteed in non-compact manifolds. However, we have the following convergence result

Theorem 1.2.1 ([96, Thm. 2.1]). *Let M be a complete connected Riemannian manifold, and $\{\Omega_i\}_{i\in\mathbb{N}}$ be a sequence of sets of volume v so that $\lim_{i\to\infty} \mathcal{P}(\Omega_i) = I_M(v)$. Then there exists a finite perimeter set Ω and sequences of finite perimeter sets $\{\Omega_i^c\}_{i\in\mathbb{N}}$, $\{\Omega_i^d\}_{i\in\mathbb{N}}$ such that:*

1. *$V(\Omega) \leqslant v$, $\mathcal{P}(\Omega) \leqslant I_M(v)$.*

2. *$V(\Omega_i^c) + V(\Omega_i^d) = v$, $\lim\limits_{i\to\infty} \mathcal{P}(\Omega_i^c) + \mathcal{P}(\Omega_i^d) = I_M(v)$.*

3. *The sequence $\{\Omega_i^d\}_{i\in\mathbb{N}}$ diverges.*

4. *Passing to a subsequence, $\lim\limits_{i\to\infty} V(\Omega_i^c) = V(\Omega)$ and $\lim\limits_{i\to\infty} \mathcal{P}(\Omega_i^c) = \mathcal{P}(\Omega)$.*

5. *Ω is an isoperimetric region for the volume it encloses.*

Roughly speaking, this result shows that a minimizing sequence has a convergent part to an isoperimetric set, perhaps with volume smaller than the original one, and a divergent part leaving any compact portion of the ambient manifold. Convergence results for minimizing sequences can be derived from Theorem 1.2.1. Morgan [66], [68, Ch. 13] used a similar idea to prove existence of clusters minimizing perimeter in Euclidean space. The author showed existence of isoperimetric regions in any complete surface with non-negative curvature [91] using a 2-dimensional version of Theorem 1.2.1. This result was also used in [96] to prove existence and non-existence results for isoperimetric regions in solid cones. Let us remark that there are examples of complete Riemannian manifolds for which isoperimetric regions do not exist for any value of the volume [90, Thm. 2.16].

Once we have obtained an isoperimetric region, a second problem is the regularity of its boundary. For this we have the following:

Theorem 1.2.2. *Let M be a smooth complete Riemannian manifold of dimension $n + 1$, and let $\Omega \subset M$ be an isoperimetric region. Then $\partial\Omega$ is smooth and has constant mean curvature except on a singular set of Hausdorff dimension less than or equal to $n - 7$.*

The proof of this theorem in Euclidean space was given by Gonzalez, Massari and Tamanini [43]; see also [67] for Riemannian manifolds with low regularity, and [68, Ch. 10].

Isoperimetric regions can be of varying topological type depending on the topological complexity of the ambient manifold. In the 3-dimensional real projective space we have the following:

Theorem 1.2.3 ([94, Thm. 8]). *Let $I_{\mathbb{S}^3}$ be the isoperimetric profile of the 3-dimensional round sphere $\mathbb{S}^3$. Then there is a value $\mu \in (0, \pi^2/2)$ so that the isoperimetric profile of $\mathbb{RP}^3$ is given by*

$$I_{\mathbb{RP}^3}(v) = \begin{cases} I_{\mathbb{S}^3}(v), & 0 \leqslant v \leqslant \mu, \\ 2v^{1/2}(\pi^2 - v^2), & \mu \leqslant v \leqslant \pi^2 - \mu, \\ I_{\mathbb{S}^3}(\pi^2 + v), & \pi^2 - \mu \leqslant v \leqslant \pi^2. \end{cases}$$

Moreover, in the first case the solution of the isoperimetric problem is a geodesic ball, in the second one it is a tubular neighborhood of a geodesic, and in the last one it is the complement of a geodesic ball.

In the Seifert manifold $\mathbb{S}^1 \times \mathbb{S}^2$ with its standard metric we have the following:

Theorem 1.2.4 ([77, Thm. 4.3]). *The isoperimetric regions in the Riemannian product $\mathbb{S}^1(r) \times \mathbb{S}^2$ are either*

1. *balls or complements of balls, or*

2. *tubular neighborhoods of the closed geodesics $\mathbb{S}^1(r) \times \{point\}$, which are diffeomorphic to $\mathbb{S}^1 \times \mathbb{S}^1$, or*

3. *sections bounded by two totally geodesic $\{point\} \times \mathbb{S}^2$, which are diffeomorphic to $[a, b] \times \mathbb{S}^2$.*

The balls are solutions for small values of the volume. If $r > 1$, then the tubes are not solutions. If $r < c_2/(\pi c_1)$, then the sections are not solutions.

Results on isoperimetric regions in quotients of $\mathbb{R}^3$ by crystallographic groups can be found in [49], [87], [88], [89], [95], and [99], amongst others.

There is a recent result by Nardulli [72], [73], [74], showing that isoperimetric solutions of small volume are invariant under the subgroup of isometries that leave invariant its center of mass.

1.3   The isoperimetric inequality in Euclidean space

Many proofs of the isoperimetric inequality (1.1) have been given using different techniques; see [19, 1.2.2]. It is usually considered that the first rigorous one in $\mathbb{R}^3$, under modern standards, was the one given by H. A. Schwarz [107]. It consisted of two steps: a symmetrization procedure (nowadays known as *Schwarz symmetriza-tion*), and a geometric construction to rule out rotationally symmetric candidates different from the sphere.

The symmetrization argument can be applied to other ambient manifolds, such as hyperbolic spaces, spheres, or to Euclidean spaces $\mathbb{R}^n$ with $O(n)$-invariant Riemannian metrics.

The geometric construction, the second step in Schwarz's argument, has to be adapted to the different ambient spaces. This was done by E. Schmidt for simply connected space forms [101], [102], [103], [105].

In this section we will prove a generalization of the Schwarz symmetrization procedure and will replace the geometric construction in Schwarz's paper by a calibration argument valid in other ambient spaces. We start with the second step.

1.3.1 A calibration argument

As a preliminary step to prove the isoperimetric inequality in Euclidean space, we give an isoperimetric inequality for sets satisfying a geometric constraint.

Theorem 1.3.1. *Let* $\Omega \subset \mathbb{R}^{n+1}$ *be a bounded closed set with piecewise smooth boundary. Let* C *be a closed right circular cylinder with axis* L *and base* D *so that*

$$D \subset \Omega \subset C. \tag{1.1}$$

Then we have

$$A(\partial\Omega) \geqslant A(\partial B), \tag{1.2}$$

where B *is the ball with* $V(B) = V(\Omega)$. *Moreover, equality holds in* (1.2) *if and only if* $\Omega = B$.

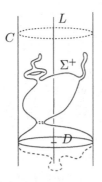

Figure 1.1: A set Σ contained in a cylinder C containing the base D of C

Proof. For simplicity, assume that L is the x_{n+1}-axis, and that D is contained in the hyperplane $x_{n+1} = 0$. In the proof we shall assume that Σ is C^1. The general case follows with minor modifications. Let S_λ be the sphere of mean curvature λ whose equatorial disc is D. Let $S_\lambda^+ = \{p \in S_\lambda : p_{n+1} \geqslant 0\}$. Translate vertically the half-sphere S_λ^+ to produce a foliation of the cylinder C. Let X be the unit normal vector (pointing downwards) to the leaves of the foliation. It is trivial to check that

$$\operatorname{div} X = -n\lambda.$$

Apply the Divergence Theorem to the vector field X in $\Omega^+ = \{p \in \Omega : p_{n+1} \geqslant 0\}$ to get

$$-n\lambda V(\Omega^+) = -\int_{\Sigma^+} \langle X, N_\Sigma \rangle \, d\Sigma - \int_D \langle X, N_D \rangle \, dD$$

$$\geqslant -A(\Sigma^+) - \int_D \langle X, N_D \rangle \, dD,$$

which implies

$$A(\Sigma^+) - n\lambda V(\Omega^+) \geqslant -\int_D \langle X, N_D \rangle \, dD.$$

Equality holds in this inequality when X and N_Σ coincide, i.e., when $\Sigma^+ = S_\lambda^+$.

Let B_λ be the ball enclosed by S_λ. The same application of the Divergence Theorem to $B_\lambda^+ = \{p \in B_\lambda : p_{n+1} \geqslant 0\}$ provides the equality

$$A(S_\lambda^+) - n\lambda V(B_\lambda^+) = -\int_D \langle X, N_D \rangle \, dD.$$

Hence we get

$$A(\Sigma^+) - n\lambda V(\Omega^+) \geqslant A(S_\lambda^+) - n\lambda V(B_\lambda^+),$$

with equality if and only if $\Sigma^+ = S_\lambda^+$.

In the same way we get

$$A(\Sigma^-) - n\lambda V(\Omega^-) \geqslant A(S_\lambda^-) - n\lambda V(B_\lambda^-),$$

with equality if and only if $\Sigma^- = B_\lambda^-$.

Adding both inequalities we get

$$A(\Sigma) - n\lambda V(\Omega) \geqslant A(S_\lambda) - n\lambda V(B_\lambda),$$

with equality if and only if $\Sigma = S_\lambda$.

Define

$$f(\mu) = A(S_\mu) - n\mu V(B_\mu) + n\mu V(\Omega),$$

where S_μ are the spheres of mean curvature μ, and B_μ are the enclosed balls. Since spheres of mean curvature μ_0 are critical points of the functional $A - n\mu_0 V$, we have

$$f'(\mu) = n\left(V(\Omega) - V(B_\mu)\right).$$

Moreover, $f''(\mu) > 0$. Hence $f(\mu)$ has a unique minimum λ_0 for which $f'(\lambda_0) = 0$ and so $V(\Omega) = V(B_{\lambda_0})$. Moreover,

$$A(\Sigma) \geqslant f(\lambda) \geqslant f(\lambda_0) = A(S_{\lambda_0}).$$

Equality holds in the above equality if and only if $\Sigma = S_\lambda$ and $\lambda = \lambda_0$. □

Theorem 1.3.1 also holds for finite perimeter sets. The proof of the inequality is straightforward. For the characterization of equality one gets that the measurable unit normal coincides with the restriction of a continuous vector field and hence the boundary is a C^1 surface by [42, Thm. 4.11].

The above result can be applied to different situations: spheres, hyperbolic spaces, Nil manifolds (Riemannian Heisenberg groups), Berger spheres. Also in spaces where regularity results for solutions of the isoperimetric problem are not yet available, like the sub-Riemannian Heisenberg groups [93].

Calibration arguments have been widely used in the calculus of variations. The reader is referred to Hélein [51], [52], to Lawlor for arguments on calibrations and slicing [57], [58], [59], and to Lawlor and Morgan [60], [61].

1.3.2 Schwarz symmetrization

Fix a line L with direction v in Euclidean space $\mathbb{R}^{n+1}$, and consider the family of hyperplanes $\Pi_t = \{p \in \mathbb{R}^{n+1} : \langle p, v \rangle = t\}$ orthogonal to L. Given $\Omega \subset \mathbb{R}^{n+1}$ with smooth boundary Σ in a generic position, Schwarz symmetrization consists in replacing the intersection $\Omega \cap \Pi_t$ by the disc in Π_t centered at $L \cap \Pi_t$ and the same n-area as $\Omega \cap \Pi_t$. In this way, a set Ω_0 is obtained so that it is rotationally symmetric with respect to L, has the same volume as Ω, and has strictly less perimeter than Ω unless Ω is rotationally symmetric with respect to a line parallel to L.

The proof of this result is based on the following, stated under much more general hypotheses.

Lemma 1.3.2. *Let M be the manifold $I \times S$, where $I \subset \mathbb{R}$ is an open interval and S is an n-dimensional Riemannian manifold with metric $d\sigma^2$. Consider on M the warped metric $ds^2 = dt^2 + f(t)^2 d\sigma^2$, where $f : I \to \mathbb{R}^+$ is a smooth function. Let $S_t = \{t\} \times S$, and let $\Omega \subset M$ be a bounded set with smooth boundary Σ. We assume that Σ is in generic position with respect to the foliation S_t, i.e., if $\rho = t|_\Sigma$, then*

$$A(\{p \in \Sigma : (\nabla_\Sigma \rho)_p = 0\}) = 0.$$

Then we have

$$A(\Sigma) \geqslant \int_I \left(\mathcal{P}(\Omega_t)^2 + \left(A(\Omega_t)' - nH_t A(\Omega_t) \right)^2 \right)^{1/2} dt, \qquad (1.3)$$

where $\Sigma_t = \Sigma \cap S_t$, $\Omega_t = \Omega \cap S_t$, and $H_t = f'(t)/f(t)$ is the mean curvature of S_t with respect to the normal $-\partial_t$. The perimeter $\mathcal{P}(\Omega_t)$ and the area $A(\Omega_t)$ are computed in S_t.

Moreover, equality holds in (1.3) *if and only if the scalar product* $\langle \partial_t, N \rangle$, *where N is the inner unit normal to Σ, is constant along Σ_t for almost all $t \in I$.*

Proof. For further reference, we define

$$\Omega_{[a,b]} = \Omega \cap t^{-1}([a,b]), \qquad \Sigma_{[a,b]} = \Sigma \cap t^{-1}([a,b]).$$

Observe that ρ is a smooth function on Σ and $(\nabla_\Sigma \rho)_p = 0$ if and only if Σ is tangent to $S_{\rho(p)}$ at p. By Sard's Theorem, for $\mathcal{L}^1$-a.e. $t \in I$, Σ is transversal to S_t.

Choose $a, b \in I$, $a < b$, so that Σ is transversal to both S_a and S_b. By the Divergence Theorem,

$$\int_{\Omega_{[a,b]}} nH_t \, dM = \int_{\Omega_{[a,b]}} \operatorname{div} \partial_t \, dM = -\int_{\Sigma_{[a,b]}} \langle \partial_t, N \rangle \, d\Sigma + A(\Omega_b) - A(\Omega_a),$$

where N is the inner unit normal to Σ. By the coarea formula, we have

$$\int_{\Omega_{[a,b]}} nH_t \, dM = \int_a^b nH_t A(\Omega_t) \, dt$$

and

$$\int_{\Sigma_{[a,b]}} \langle \partial_t, N \rangle \, d\Sigma = \int_a^b \left\{ \int_{\Sigma_t} \frac{\langle \partial_t, N \rangle}{|\nabla_\Sigma \rho|} \, d\Sigma_t \right\} dt.$$

So we obtain

$$\left. \frac{d}{dt} \right|_{t=a} A(\Omega_t) = nH_a A(\Omega_a) + \int_{\Sigma_a} \frac{\langle \partial_t, N \rangle}{|\nabla_\Sigma \rho|} \, d\Sigma_a, \tag{1.4}$$

for any regular value a of ρ.

Now, again by the coarea formula, using Minkowski's inequality [47, §6.13] in the third step, we get

$$A(\Sigma) = \int_I \left\{ \int_{\Sigma_t} \frac{1}{|\nabla_\Sigma \rho|} \, d\Sigma_t \right\} dt$$

$$= \int_I \left\{ \int_{\Sigma_t} \left(1 + \frac{1 - |\nabla_\Sigma \rho|^2}{|\nabla_\Sigma \rho|^2} \right)^{1/2} d\Sigma_t \right\} dt$$

$$\geqslant \left(\left(\int_{\Sigma_t} d\Sigma_t \right)^2 + \left(\int_{\Sigma_t} \frac{\sqrt{1 - |\nabla_\Sigma \rho|^2}}{|\nabla_\Sigma \rho|} \, d\Sigma_t \right)^2 \right)^{1/2}$$

$$= \left(P(\Omega_t)^2 + \left(\int_{\Sigma_t} \frac{\sqrt{1 - |\nabla_\Sigma \rho|^2}}{|\nabla_\Sigma \rho|} \, d\Sigma_t \right)^2 \right)^{1/2}, \tag{1.5}$$

with equality if and only if $\sqrt{1 - |\nabla_\Sigma \rho|^2}$ is constant along Σ_t. Observe that, for a regular value $t \in I$ of ρ, we have

$$\langle \partial_t, N \rangle^2 = 1 - |\nabla_\Sigma \rho|^2,$$

and so

$$\left(\int_{\Sigma_t} \frac{\sqrt{1 - |\nabla_\Sigma \rho|^2}}{|\nabla_\Sigma \rho|} \, d\Sigma_t \right)^2 = \left(\int_{\Sigma_t} \frac{\langle \partial_t, N \rangle}{|\nabla_\Sigma \rho|} \, d\Sigma_t \right)^2.$$

Now equations (1.4) and (1.5) imply (1.3). Equality is easily characterized. $\qquad\square$

This result can be applied in many different situations. For instance one can consider, in $\mathbb{R}^{n+1}$, the function $f(x_1, \ldots, x_{n+1}) = x_{n+1}$. In this case, the hypersurfaces S_t are the hyperplanes parallel to $x_{n+1} = 0$, and the symmetrization in $\mathbb{R}^{n+1}$ can be proved by induction once isoperimetric solutions in $\mathbb{R}^n$ have been found. This is the classical Schwarz symmetrization. In case S_t are the balls centered at a given point, we get the spherical symmetrization considered by Fred Almgren [5].

To simplify the application of Schwarz symmetrization in $\mathbb{R}^{n+1}$, we need to choose an appropriate height function. Given $v \in \mathbb{S}^n$, we define the *height function* associated to v by

$$h_v(p) = \langle p, v \rangle.$$

Fix a compact embedded hypersurface $\Sigma \subset \mathbb{R}^{n+1}$. For $v \in \mathbb{S}^n$, define $\rho_v = h_v|_\Sigma$. It is immediate to check that $p \in \Sigma$ is a critical point of ρ_v if and only if $N_p = \pm v$, where N is the inner unit normal to Σ. On the other hand, the Hessian $\nabla_\Sigma^2 \rho_v$ of ρ_v in Σ in a critical point p of ρ_v can be easily computed. If $v_1, v_2 \in T_p\Sigma$, then

$$\nabla_\Sigma^2 \rho_v(v_1, v_2) = \langle N, v \rangle \, \sigma(v_1, v_2) = \pm \sigma(v_1, v_2),$$

where σ is the second fundamental form of Σ. Hence $p \in \Sigma$ is a non-degenerate critical point of ρ_v if and only if $N_p = \pm v$ and the principal curvatures of Σ at p are all different from zero.

Consider now the normal Gauss map $\varphi \colon \Sigma \to \mathbb{S}^n$. The Jacobian of φ at $p \in \Sigma$ equals the Gauss–Kronecker curvature of Σ, i.e., the product of the principal curvatures of Σ. By Sard's Theorem, the set of critical values of φ has $\mathcal{H}^n$-measure 0. Hence, for $\mathcal{H}^n$-a.e. $v \in \mathbb{S}^n$, the points in Σ with normal $\pm v$ are non-degenerate critical points of ρ_v. This implies that ρ_v is a Morse function on Σ. In particular, there is only a finite number of critical points by the compactness of Σ.

Now we are ready to apply Schwarz symmetrization

Theorem 1.3.3. *Let $\Omega \subset \mathbb{R}^{n+1}$ be a bounded open set with smooth boundary Σ. Let $v \in \mathbb{S}^n$ so that ρ_v, the restriction to Σ of the height function defined by v, is a Morse function. Let L be a fixed line with direction v. Let $S_t = \{p \in \mathbb{R}^{n+1} : \langle p, v \rangle = t\}$, $\Omega_t = \Omega \cap S_t$, $\Sigma_t = \Sigma \cap S_t$. For every $t \in \mathbb{R}$ such that $\Omega_t \neq \emptyset$, consider the disc Ω_t^* with the same area as Ω_t centered at $L \cap S_t$. Let*

$$\Omega = \bigcup_{t \in \mathbb{R}} \Omega_t^*, \qquad \Sigma^* = \partial \Omega^*.$$

Then:

1. $V(\Omega^*) = V(\Omega)$.

2. Σ^* is a piecewise smooth surface with $A(\Sigma) \geqslant A(\Sigma^*)$. Moreover, equality holds if and only if Ω is a set of revolution around some line parallel to L. In this case, Σ^* is a translation of Σ.

3. $A(\Sigma) \geqslant A(\partial B)$, where $B \subset \mathbb{R}^{n+1}$ is a ball with $V(B) = V(\Omega)$. Moreover, equality holds if and only if Ω is a ball.

Proof. In all cases, the equality $V(\Omega^*) = V(\Omega)$ is a trivial consequence of the coarea formula.

We shall consider first the planar case with some detail. The coordinates in $\mathbb{R}^2$ will be denoted by (x, y) instead of (x_1, x_2). After a rotation if necessary, we may assume that $v = (0, 1)$, so that ρ_v is the restriction of y to the boundary curve Σ. Let $y_1 < \cdots < y_k$ be the critical values of ρ_v. For any $y_0 \in (y_i, y_{i+1})$, $i = 1, \ldots, k - 1$ the number of connected components of the intersection of the line $y = y_0$ with Ω is a constant m, and there exist smooth functions

$$f_1 < g_1 < f_2 < g_2 < \cdots < f_m < g_m,$$

so that $(x, y) \in \Omega \cap \rho_v^{-1}(y_i, y_{i+1})$ if and only if $x \in \bigcup_{i=1}^m (f_j(y), g_j(y))$. It follows that $(x, y) \in \Omega^* \cap \rho_v^{-1}(y_i, y_{i+1})$ if and only if

$$x \in (-f(y), f(y)),$$

where

$$f(y) = \frac{1}{2} \sum_{j=1}^m (g_j(y) - f_j(y)). \tag{1.6}$$

The length of $\Sigma \cap \rho_v^{-1}(y_i, y_{i+1})$ is given by

$$L(\Sigma \cap \rho_v^{-1}(y_i, y_{i+1})) = \int_{y_i}^{y_{i+1}} \sum_{j=1}^m \left\{ \sqrt{1 + g_j'(y)^2} + \sqrt{1 + f_j'(y)^2} \right\} dy,$$

and the one of $\Sigma^* \cap \rho_v^{-1}(y_i, y_{i+1})$ by

$$L(\Sigma^* \cap \rho_v^{-1}(y_i, y_{i+1})) = 2 \int_{y_i}^{y_{i+1}} \sqrt{1 + f'(y)^2}\, dy.$$

By (1.6), we have

$$(2m, 2f'(y)) = \sum_{j=1}^m (1, g_j'(y)) - (-1, f_j'(y)),$$

so that, by the triangle inequality,

$$2\sqrt{1 + f'(y)^2} \leqslant 2\sqrt{m^2 + f'(y)^2} \leqslant \sum_{j=1}^m \left\{ \sqrt{1 + g_j'(y)^2} + \sqrt{1 + f_j'(y)^2} \right\} \tag{1.7}$$

which implies $L(\Sigma^* \cap \rho_v^{-1}(y_i, y_{i+1})) \leqslant L(\Sigma \cap \rho_v^{-1}(y_i, y_{i+1}))$. If equality holds in (1.7), then $m = 1$ and $g_1 + f_1$ is a constant function in (y_i, y_{i+1}). This implies that $\Sigma \cap \rho_v^{-1}(y_i, y_{i+1})$ is symmetric with respect to some vertical line.

Applying these arguments to each interval (y_i, y_{i+1}), we infer that $L(\Sigma^*) \leqslant L(\Sigma)$, and that equality holds if and only if Σ is symmetric with respect to a vertical line. Now we apply Theorem 1.3.1 to obtain $L(\Sigma^*) \geqslant L(\partial D)$, where D is the planar disc satisfying $A(D) = A(\Omega^*) = A(\Omega)$. Hence $L(\partial D) \leqslant L(\Sigma^*) \leqslant L(\Sigma)$. In case equality holds, then $\Sigma^* = D$ by Theorem 1.3.1 and so Σ is a disc.

Now we treat the higher dimensional case. The proof will follow by induction on the dimension $n + 1$ of the Euclidean space. The case $n + 1 = 2$ has already been considered, so we assume that $n + 1 \geqslant 3$. Assume that $v = (0, \dots, 0, 1)$.

By construction, Ω^* is a set of revolution with smooth boundary out of the critical values of ρ_v. By the induction hypothesis, we have $\mathcal{P}(\Omega_t) \geqslant \mathcal{P}(\Omega_t^*)$. As ρ_v is a Morse function, Σ^* is piecewise smooth and we can apply Lemma 1.3.2 to conclude that

$$A(\Sigma) \geqslant A(\Sigma^*).$$

If equality holds in the above inequality, then Σ_t is a disc for all t, and $\langle \partial_t, N \rangle$ is constant along Σ_t for all t. It is straightforward to check that this implies that Ω is a set of revolution around a line parallel to L. This proves 2.

Now 3 follows from 2 and Theorem 1.3.1. □

We would like to remark that there are trivial proofs of the planar isoperimetric inequality, such as the one due to Hurwitz; see [25, Thm. V.5.1].

There are examples of sets for which Schwarz symmetrization does not decrease the boundary area. They are easily constructed from hypersurfaces with intersection of positive area with one of the leaves S_t. Of course, in these cases, the relevant height function is not a Morse function.

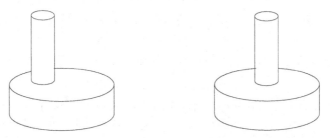

Figure 1.2: An example of perimeter non-decreasing under Schwarz symmetrization. Examples with smooth boundary can be obtained by rounding the corners of the above surfaces

Apart from Schwarz's symmetrization, some others can be used in different problems, like Steiner's, in which a hyperplane is chosen, and the intersections of a set with lines orthogonal to the hyperplane are replaced by segments in the same line, of the same one-dimensional Hausdorff measure, but centered at a point of the hyperplane, [19, §9.2]. A powerful symmetrization procedure in Riemannian geometry was introduced by W. Y. Hsiang [53], [54].

1.3.3   Another proof of the isoperimetric inequality

A somewhat different proof of the isoperimetric inequality in Euclidean space was given by Montiel and Ros [65]. Note that isoperimetric regions always exist in Euclidean spaces, since the quotient by the isometry group is compact [68, pp. 128–129], and their boundaries are smooth hypersurfaces with constant mean curvature except on a singular set of Hausdorff dimension less than or equal to $n - 7$ (the dimension of the ambient manifold is assumed to be $n + 1$). The result of Montiel and Ros, using ideas from Gromov [46, App. C$_+$], and Heintze and Karcher [50], is the following.

Theorem 1.3.4 ([65]). *Let $\Sigma \subset \mathbb{R}^{n+1}$ be a compact smooth embedded hypersurface with constant mean curvature H, enclosing a region Ω. Then*

$$(n + 1) H V(\Omega) \leqslant A(\Sigma). \tag{1.8}$$

Equality holds if and only if Ω is a geodesic ball (and Σ is a metric sphere).

Proof. Let N be the inner normal to Σ. We recall that the mean curvature is the arithmetic mean of the principal curvatures. For every $p \in \Sigma$, let $c(p)$ be the first focal point along the geodesic normal to Σ passing through p. As Σ is compact, $H > 0$ by the maximum principle for the mean curvature equation. Let $\kappa^+(p)$ be the maximum of the positive principal curvatures of Σ at p. Trivially we have $\kappa^+ \geqslant H$. Then $c(p) \leqslant 1/\kappa^+ \leqslant 1/H$. So we have

$$V(\Omega) \leqslant \int_0^{c(p)} \left\{ \int_\Sigma \prod_{i=1}^n (1 - t\kappa_i(p)) \, d\Sigma(p) \right\} dt$$

$$\leqslant \int_\Sigma \left\{ \int_0^{1/H} (1 - tH)^n \, dt \right\} d\Sigma$$

$$= \frac{1}{(n + 1) H} A(\Sigma),$$

where in the second inequality we have used the arithmetic-geometric inequality $(a_1 \cdots a_n)^{1/n} \leqslant (a_1 + \cdots + a_n)/n$, for $a_i \geqslant 0$, with equality if and only if $a_1 = \cdots = a_n$. This proves (1.8).

In case equality holds in (1.8), then $c(p) = 1/H$ for all p and Σ is totally umbilical. This implies that Σ is an umbilical sphere and Ω is a ball.   □

As a corollary, we have:

Theorem 1.3.5 (Alexandrov Theorem in Euclidean space). *Let $\Sigma \subset \mathbb{R}^{n+1}$ be a smooth compact embedded hypersurface with constant mean curvature. Then Σ is a round sphere.*

Proof. Observe that Σ is a critical point of the functional $A - nHV$. Consider the vector field

$$X = \sum_{i=1}^{n+1} x_i \frac{\partial}{\partial x_i}.$$

Then $\nabla_u X = u$ for all $p \in \mathbb{R}^{n+1}$ and $u \in T_p\mathbb{R}^{n+1}$. This implies

$$\operatorname{div} X = n + 1, \qquad \operatorname{div}_\Sigma X = n.$$

Using the first variation formulas for the area and the volume, and the fact that Σ is critical for $A - nHV$, we get

$$0 = nA - nH(n+1)V = n\left(A - (n+1)HV\right).$$

Then, by Theorem 1.3.4, Σ is a round sphere. $\qquad\square$

The same proof holds when Σ is the boundary of an isoperimetric region. Observe that, for every q in the interior of Ω, the points in Σ at minimum distance from q are regular points (since the minimal tangent cone of $\mathbb{R}^{n+1} - \Omega$ at those points is contained in a halfspace). Look at Gromov's proof of the isoperimetric inequality in manifolds with lower bounds on the Ricci curvature [46, App. C$_+$].

Theorem 1.3.4 was later generalized by Montiel [64] to certain manifolds of dimension larger than or equal to 3 admitting a conformal vector field.

The classical proof that the only compact embedded hypersurface with constant mean curvature in $\mathbb{R}^{n+1}$ is the round sphere was proved by Alexandrov using a reflection-moving planes argument [3].

Chapter 2

Surfaces

2.1 Introduction

In the Euclidean plane, the length L of a closed embedded curve, not necessarily connected, and the enclosed area A, satisfy the *isoperimetric inequality*

$$L^2 \geqslant 4\pi A, \tag{2.1}$$

with equality if and only if the curve is a circle. The arguments in the previous chapter, using Steiner–Schwarz symmetrization with respect to a chosen line, and a calibration argument, imply inequality (2.1). Many different proofs of (2.1) have been given using a wide variety of techniques (see [19, Ch. 1], [24], [25, Ch. V], and [75]). Inequality (2.1) has been extended in several ways. A useful generalization is the following: Assume that M is a simply connected Riemannian surface with Gauss curvature K bounded from above by a constant K_0. Then

$$L^2 \geqslant 4\pi A - K_0 A^2, \tag{2.2}$$

and equality holds only for geodesic circles enclosing a disc of constant curvature K_0; see [19, Ch. 1]. A proof of (2.2) using inner parallels when the Riemannian metric is analytic and the boundary curve is analytic was given by Bol [17] and Fiala [35]; see also [25, Ch. V.5]. An extension by approximation to general metrics was given by Chavel and Feldman [26]. A direct proof for smooth metrics was given in Hartman [48]; see also [108] and [109]. An excellent survey with an exhaustive list of references is the one by R. Osserman [75].

Inequality (2.2) is known as *Bol–Fiala's inequality* and is a particular case of the inequality

$$L^2 + 2\left(\omega_{K_0}^+ - 2\pi\chi\right)A + K_0 A^2 \geqslant 0;$$

see [19, Thm. 2.2.1]. Here L and A are the perimeter and area of a domain Ω with compact closure in a Riemannian surface M such that $\partial\Omega$ consists of a finite number of rectifiable curves, χ is the Euler characteristic of Ω, and

$$w^+_{K_0} = \int_\Omega (K - K_0)^+ \, dM.$$

Until recently, the isoperimetric regions were not known in surfaces of variable curvature. In 1996, Benjamini and Cao [15] characterized the isoperimetric regions in planes of revolution with non-increasing Gauss curvature as a function of the distance to a given point, including the paraboloid of revolution $z = x^2 + y^2$. Their proof used the result by Grayson [45] on the convergence of the curve shortening flow.

2.1.1   The Gauss–Bonnet Theorem

One of the main tools when using a deformation approach to obtain isoperimetric inequalities, either by inner parallels or by the curve shortening flow, is the Gauss–Bonnet Theorem. To keep these notes self-contained, we briefly recall Blaschke's variational proof of this result.

Theorem 2.1.1 (Gauss–Bonnet Theorem [16]). *Let M be a Riemannian surface, and $\Omega \subset M$ a disc with smooth boundary C. Let K be the Gauss curvature of M and h the geodesic curvature of C with respect to the inner unit normal N. Then we have*

$$\int_\Omega K \, dM + \int_C h \, dC = 2\pi. \tag{2.3}$$

Proof. Let U be any vector field with compact support in M, and let $u = \langle U, N \rangle$. Consider the flow $\{\varphi_t\}_{t\in\mathbb{R}}$ associated to U, and let $\Omega_t = \varphi_t(\Omega)$, $C_t = \varphi_t(C)$. Then the variation of the total Gauss curvature of Ω under the action of this flow is given by

$$\frac{d}{dt}\Big|_{t=0} \int_{\Omega_t} K \, dM = \int_\Omega \left(U(K) + K \operatorname{div} U \right) dM$$

$$= \int_\Omega \operatorname{div}(KU) \, dM = - \int_C Ku \, dC.$$

On the other hand, the variation of the integral of the geodesic curvature is given by

$$\frac{d}{dt}\Big|_{t=0} \int_{C_t} h_t \, dC_t = \int_C \left(U(h_t) + h \operatorname{div}_C U \right) dC$$

$$= \int_C \left(U^\top(h) + U^\perp(h_t) + h \operatorname{div}_C U^\top + h \operatorname{div}_C U^\perp \right) dC$$

$$= \int_C \left(\operatorname{div}_C(hU^\top) + U^\perp(h_t) + h \operatorname{div}_C U^\perp \right) dC.$$

By the Divergence Theorem,

$$\int_C \operatorname{div}_C (hU^\top)\, dC = 0.$$

On the other hand, $\operatorname{div}_C U^\perp = -hu$, and the derivative of the geodesic curvature can be computed as $U^\perp(h_t) = u'' + (K + h^2)\, u$; see [11], where the prime is the derivative with respect to arc-length in C. So we have

$$\int_C U^\perp(h_t)\, dC = \int_C (K + h^2)\, u\, dC,$$

and we finally get

$$\frac{d}{dt}\bigg|_{t=0} \left(\int_{\Omega_t} K\, dM + \int_{C_t} h_t\, dC_t \right) = 0.$$

Hence the left side of (2.3) is invariant under the action of the flow φ_t associated to U. Contracting the curve C to a point in the interior of Ω so that the curves approaching the point are geodesic circles, we get that the constant value of the left side of (2.3) is equal to 2π, which yields the desired result. $\qquad\square$

From formula (2.3) one can derive the classical Gauss–Bonnet formula for piecewise smooth curves [25, Thm. V.2.5] and the Gauss–Bonnet Theorem for closed Riemannian surfaces [25, Thm. V.2.3]. The reader is referred to the clear exposition of Chavel [25, §V.2] for a rather complete description of these consequences of (2.3).

The Gauss–Bonnet formula (2.3) allows control of the integral $\int_C h\, dC$, which is in fact the derivative of the length when we take inner parallels to the curve C. Then one can prove the isoperimetric inequality (2.1) in the following way: consider a domain Ω bounded by a Jordan curve C in a Riemannian surface with $K \leqslant K_0$, and the inner parallel regions $\Omega_t = \{p \in \Omega : d(p, C) \geqslant t\}$, which are bounded by the sets $C_t = \{p \in \Omega : d(p, C) = t\}$. At least while the sets C_t are smooth embedded curves (and we know that this happens when t is small enough if C is at least C^2), the derivatives of area and length are given, respectively, by

$$\frac{dA(\Omega_t)}{dt} = -L(C_t),$$

$$\frac{dL(C_t)}{dt} = -\int_{C_t} h_t\, dC_t = -2\pi + \int_{\Omega_t} K\, dM.$$

Hence we can take the area as a parameter for the deformation, and

$$\frac{dL^2}{dA} = 2\left(2\pi - \int_{\Omega_t} K\, dA \right) \geqslant 2\, (2\pi - K_0\, A).$$

In case the deformation takes the regions Ω_t to some set of area zero, we merely integrate the above inequality to obtain (2.2). Usually this deformation by inner parallels is going to develop singularities, so the task is to control the behavior of length and area when crossing these singular times. In the analytic case, both $A(t)$ and $L(t)$ are continuous functions of t and the singularities of these deformations were studied by Bol [17] and Fiala [35]; see also [25, §V.5, pp. 255–263]. In the non-analytic case, the length function $L(t)$ may be discontinuous [48], and the study of their evolution is much more involved.

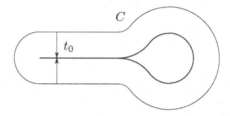

Figure 2.1: A typical discontinuity of $L(t)$ at $t = t_0$

2.2 Curve shortening flow

When using inner parallels to prove isoperimetric inequalities, one has to deal with the singularities developed along the deformation. One can try to replace this deformation by any other one so that a good control of the derivatives of length and area is kept, while avoiding singularity formation. A natural candidate is the curve shortening flow.

2.2.1 Basic results

Given a Jordan curve $\alpha \colon \mathbb{S}^1 \to M$ with inner normal n, we deform it by the geodesic curvature flow

$$\frac{\partial \alpha_t}{\partial t} = h_t n_t, \tag{2.1}$$

where h_t denotes the geodesic curvature with respect to the inner unit normal vector field n_t to α_t. A fundamental problem is to understand the evolution equation (2.1) for large values of t. Solutions of (2.1) are defined for $t \in [0, t_\infty)$, but not on a larger interval. The value t_∞ is called the *maximal time* for the solution $\alpha_t \colon \mathbb{S}^1 \times [0, t_\infty) \to M$. After several results by Angenent [6], Gage [38], [39], Gage–Hamilton [37], Abresch–Langer [1], Grayson [44], the main result in this theory was proved by Grayson [45]; see Theorem 2.2.1 below. The *convex hull* of a set $A \subset M$ is the smallest convex set containing A. We shall say that M is *convex at infinity* if the convex hull of every compact set is compact. Since an evolving curve cannot leave a locally convex region by the avoidance principle in subsection 2.2.2, convexity at infinity is a hypothesis to ensure that the flow is confined to a compact region.

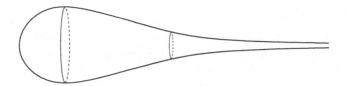

Figure 2.2: A surface of revolution which is not convex at infinity

Theorem 2.2.1 ([45]). *Let M be a complete surface which is convex at infinity, α a Jordan curve in M, and α_t a solution of the curve shortening flow equation (2.1). Let t_∞ be the maximal time for this solution. Then:*

1. *α_t remains embedded for all t.*

2. *If $t_\infty < +\infty$, then α_t converges to a point as $t \to t_\infty$.*

3. *If $t_\infty = +\infty$, then α_t converges to a closed geodesic as $t \to t_\infty$.*

Assume that the geodesic curvature flow is applied to a Jordan curve $C = \alpha(\mathbb{S}^1)$ enclosing a region Ω. Denote by C_t the Jordan curve $\alpha_t(C)$, and let Ω_t be the region enclosed by C_t. Let $A(t)$ be the area of the region Ω_t and $L(t)$ be the length of C_t. The variation formula for the area and the Gauss–Bonnet Theorem imply

$$\frac{dA}{dt} = -\int_{C_t} h_t \, dC_t = -2\pi + \int_{\Omega_t} K \, dM \leqslant -2\pi + \int_M K^+ \, dM. \qquad (2.2)$$

Hence geometric conditions must be imposed on M to guarantee that $dA/dt < 0$. The derivative of length along the curve shortening flow is given by

$$\frac{dL}{dt} = -\int_{C_t} h_t^2 \, dC_t,$$

which is strictly negative whenever $h_t \not\equiv 0$. Assuming that the area is strictly decreasing along the curve shortening flow, we have

$$\frac{dL^2}{dA} = 2L \frac{\int_{C_t} h_t^2 \, dC_t}{\int_{C_t} h_t \, dC_t} \geqslant 2\int_{C_t} h_t \, dC_t, \qquad (2.3)$$

by the Schwarz inequality $\left(\int_{C_t} h_t \, dC_t\right)^2 \leqslant L(t) \int_{C_t} h_t^2 \, dC_t$. Hence we get from (2.3)

$$\frac{dL^2}{dA} \geqslant 2\left(2\pi - \int_{\Omega_t} K \, dA\right),$$

which is analogous to the formula obtained when taking inner parallels to the curve C.

2.2.2   The avoidance principle

A fundamental property of the curve shortening flow is the avoidance principle, which asserts that two disjoint curves remain disjoint under the action of the flow. The avoidance principle is based on the maximum principle for parabolic linear equations.

Theorem 2.2.2 ([86, Lemma 3]). *Let L be the differential operator*

$$L(u) \equiv a(x,t)\frac{\partial^2 u}{\partial x^2} + b(x,t)\frac{\partial u}{\partial x} - \frac{\partial u}{\partial t}, \tag{2.4}$$

where a and b are smooth and bounded and L is uniformly parabolic. Suppose that, in a domain E of the plane xt, u satisfies the inequality $L(u) \geq 0$. Assume that $u < M$ in the portion of E lying in the strip $t_0 < t < t_1$ for some fixed numbers t_0 and t_1. Then $u < M$ on the portion of the line $t = t_1$ contained in E.

Of course, if the function u is a solution of the linear parabolic equation

$$\frac{\partial u}{\partial t} = a\frac{\partial^2 u}{\partial x^2} + b\frac{\partial u}{\partial x} + cu,$$

with $cu \leq 0$, then u satisfies $L(u) \geq 0$.

Now we can prove the avoidance principle for the curve shortening flow

Theorem 2.2.3 (Avoidance principle [31, Prop. 1.6]). *Let C_1, $C_2 \subset M$ be disjoint Jordan curves in a complete Riemannian surface. Let $(C_1)_t$, $(C_2)_t$ be solutions of the curve shortening flow defined in maximal time intervals $[0, T_1)$, $[0, T_2)$, respectively. Then $(C_1)_t$, $(C_2)_t$ are disjoint for all $t < \min\{T_1, T_2\}$.*

Proof. Assume $\bar{t} < \min\{T_1, T_2\}$ is the first contact time. Let $p \in (C_1)_{\bar{t}} \cap (C_2)_{\bar{t}}$, and let Γ be a curve tangent to both $(C_1)_{\bar{t}}$ and $(C_2)_{\bar{t}}$ at p. Consider Fermi coordinates (x, y), [25, p. 142], based on the curve $\Gamma = \Gamma(x)$.

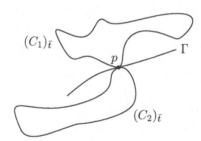

Figure 2.3: The first contact time $\bar{t}$

Then the matrix of the Riemannian metric g in coordinates (x, y) is given by

$$\begin{pmatrix} g_{11} & 0 \\ 0 & 1 \end{pmatrix}.$$

The curves $(C_1)_{\bar{t}}$, $(C_2)_{\bar{t}}$ are represented near $(\bar{x}, \bar{t})$, where $\bar{x}$ satisfies $\Gamma(\bar{x}) = p$, by the graphs of the functions $u_1(x, t)$, $u_2(x, t)$, respectively, and they satisfy the evolution equation

$$\frac{\partial u_i}{\partial t} = \frac{1}{(g_{11} + (u_i)_x^2)} (u_i)_{xx} + \tilde{\Phi}(x, t, u_i, (u_i)_x), \qquad i = 1, 2.$$

Assume that their difference $u = u_1 - u_2$ is strictly negative for $t < \bar{t}$. Then u satisfies a linear parabolic equation

$$\frac{\partial u}{\partial t} = a u_{xx} + b u_x + c u.$$

In order to apply the maximum principle for parabolic equations we need $c \leqslant 0$, which does not hold in general. So instead of u we consider the function $w(x, t) = e^{-\lambda t} u(x, t)$, and we have

$$\frac{\partial w}{\partial t} = -\lambda w + e^{-\lambda t} \left(a u_{xx} + b u_x + c u \right)$$

$$= a w_{xx} + b w_x + (c - \lambda) w,$$

which satisfies $c - \lambda < 0$ for λ sufficiently large. From Theorem 2.2.2 we get a contradiction. $\qquad\square$

2.3 Applications of curve shortening flow to isoperimetric inequalities

2.3.1 Planes

Using Grayson's Theorem, Benjamini and Cao [15] were able to prove the following isoperimetric result.

Theorem 2.3.1 ([15]). *Let M be a complete plane of revolution about a given vertex $o \in M$ so that the Gauss curvature is a non-increasing function of the distance to o. Let B_r be the geodesic ball of radius $r > 0$ centered at o, and assume that*

$$\int_{B_r} K^+ \, dM < 2\pi, \qquad \text{for all } r > 0. \tag{2.1}$$

Then, given a relatively compact set $\Omega \subset\subset M$ with smooth boundary we have

$$L(\partial\Omega) \geqslant L(\partial B_r), \tag{2.2}$$

where B_r is the ball with $A(B_r) = A(\Omega)$. Equality holds in (2.2) if and only if Ω is isometric to B_r.

This result can be applied to the paraboloid of revolution $z = x^2 + y^2$. Geodesic circles centered at o have constant geodesic curvature which is positive by the Gauss–Bonnet Theorem and (2.1), and so M is convex at infinity. The Gauss–Bonnet Theorem implies also that there are no closed geodesics in M, so the curve shortening flow deforms any Jordan curve to a point.

We shall give a simplified proof of Theorem 2.3.1 using the avoidance principle. We have also included the original proof by Benjamini and Cao because of its geometrical interest.

Proof of Theorem 2.3.1 using the avoidance principle. Prior to the proof we make the following observations:

1. There are no closed geodesics in M by the Gauss–Bonnet Theorem.

2. Along the curve shortening flow the area decreases by equation (2.2) and the condition $\int_{B_r} K^+ \, dM < 2\pi$.

3. The curve shortening flow applied to circles centered at o provides a family of circles contracting to o.

4. M is convex at infinity since geodesic balls are convex.

We shall assume first that Ω is a disc. The curve shortening flow applied to $C = \partial\Omega$ collapses the curve to a point, since there are no closed geodesics in M. Hence the area can be taken as a parameter of the deformation and we have

$$\frac{dL^2}{dA} = \frac{2L \int_{C_t} h_t^2 \, dC_t}{\int_{C_t} h_t \, dC_t} \geqslant 2 \int_{C_t} h_t \, dC_t = 2\left(2\pi - \int_{\Omega_A} K \, dM\right)$$

$$\geqslant 2\left(2\pi - \int_{B_{r(A)}} K \, dM\right) = \frac{dL^2}{dA}(B_{r(A)}),$$

where Ω_A is the disc enclosed by some C_t of area $A \leqslant A(\Omega)$. In the first line of the above displayed inequality we have used the Schwarz inequality and, in the second one, the fact that the balls centered at o maximize the total curvature for any given area. Integrating the above inequality we get (2.2). In case equality holds we get that h_t is constant along the deformation and that $\Omega_A = B_{r(A)}$ for all $A \in (0, A(\Omega)]$.

Consider now the case in which Ω is connected but the Euler characteristic $\chi(\Omega) \leqslant 0$. The set Ω is a disc D from which a finite number of discs have been removed. We trivially have

$$L(\partial\Omega) > L(\partial D), \qquad A(\Omega) < A(D).$$

Apply the curve shortening flow to D. Let t_0 be the instant for which $A(D_{t_0}) = A(\Omega)$. From the previous case we get

$$L(\partial D_{t_0}) \geqslant L(\partial B_{r(A(D_{t_0}))}),$$

and we finally get

$$L(\partial\Omega) > L(\partial D) \geqslant L(\partial D_{t_0}) \geqslant L(\partial B_{r(A(\Omega))}).$$

Observe that in this case the equality is never attained.

As a third case, assume that $\Omega = \Omega_1 \cup \cdots \cup \Omega_k$, where the sets Ω_i are disjoint discs. Let $T_1 \geqslant \cdots \geqslant T_k$ be the maximal times for the flows $(\partial\Omega_1)_t, \ldots, (\partial\Omega_k)_t$, and define

$$\Omega_t = \begin{cases} (\Omega_1)_t \cup \cdots \cup (\Omega_{k-1})_t \cup (\Omega_k)_t, & 0 \leqslant t \leqslant T_k, \\ (\Omega_1)_t \cup \cdots \cup (\Omega_{k-1})_t, & T_k \leqslant t \leqslant T_{k-1}, \\ \cdots & \cdots \\ (\Omega_1)_t, & T_2 \leqslant t \leqslant T_1. \end{cases}$$

Observe that, by the avoidance principle, the sets $(\Omega_1)_t, \ldots, (\Omega_k)_t$ are disjoint whenever they exist. Let $C_t = \partial\Omega_t$. Now we have, on the interval $[T_i, T_{i-1}]$,

$$\frac{dL^2}{dA} = \frac{2L \int_{C_t} h_t^2 \, dC_t}{\int_{C_t} h_t \, dC_t} \geqslant 2 \int_{C_t} h_t \, dC_t = 2 \left(2\pi \, (i-1) - \int_{\Omega_t} K \, dM \right)$$

$$\geqslant 2 \left(2\pi - \int_{\Omega_t} K \, dM \right),$$

and the last inequality is strict if $i > 2$. Now we can argue as in the case of a single disc to conclude that

$$L(\partial\Omega) \geqslant L(\partial B_{r(A)}).$$

Moreover, the inequality is strict if Ω has more than one component.

Finally assume that $\Omega = \Omega_1 \cup \cdots \cup \Omega_k$, where the sets Ω_i are disjoint. Each Ω_i is a disc D_i from which a finite number of discs have been removed. The discs D_i are either disjoint or nested. We say that D_i is *outermost* if it is not contained in another D_j. Let $\mathcal{J}$ be the family of indexes corresponding to outermost discs, and

$$D = \bigcup_{i \in \mathcal{J}} D_i.$$

Then we have

$$L(\partial\Omega) > L(\partial D), \qquad A(\Omega) < A(D),$$

and we may apply the curve shortening flow to the union of discs D as in the previous case to conclude that $L(\partial\Omega) > L(\partial B_{r(A)})$. $\square$

Proof of Theorem 2.3.1 following Benjamini and Cao. Assume first that Ω is a domain with $\chi(\Omega) \leqslant 0$. As M is a plane, Ω is obtained from a disc D deleting a finite number of discs with smooth boundary. Obviously, $A(\Omega) < A(D)$ and $L(\partial\Omega) > L(\partial D)$. Apply the geodesic curvature flow to ∂D and stop at some

curve C enclosing a region Ω^* with $A(\Omega^*) = A(\Omega)$. Obviously, $L(\partial\Omega^*) \leqslant L(\partial D) <$
$L(\partial\Omega)$. Hence Ω with $\chi(\Omega) \leqslant 0$ cannot be an isoperimetric region.

Assume now that Ω is a disc. One can use the geodesic curvature flow to
obtain a family of domains Ω_a, for $a \in (0, A(\Omega))$. Let $r(a)$ be the radius of the
geodesic ball centered at o of area a. Since ∂B_r has constant geodesic curvature
for all r, we get that the geodesic curvature flow applied to a circle centered at o
yields a family of contracting circles centered at o. By (2.3) and Gauss–Bonnet,
we have

$$\frac{dL^2(\partial\Omega_a)}{da} \geqslant 2\left\{2\pi - \int_{\Omega_a} K\,dM\right\}$$

$$\geqslant 2\left\{2\pi - \int_{B_{r(a)}} K\,dM\right\} = \frac{dL^2(\partial B_{r(a)})}{da},$$

since K is non-increasing. Since $\liminf_{a\to 0} L(\partial\Omega_a) \geqslant 0 = \lim_{a\to 0} L(\partial B_{r(a)})$, we
get

$$L(\partial\Omega) = L(\partial\Omega_{A(\Omega)}) \geqslant L(\partial B_{r(A(\Omega))}).$$

In case $L(\partial\Omega) = L(\partial B_{r(A(\Omega))})$, we get

$$\int_{\Omega_a} K\,dM = \int_{B_{r(a)}} K\,dM,$$

for all $a \in (0, A(\Omega))$. Since K is non-increasing we easily conclude that Ω is
isometric to $B_{r(A(\Omega))}$.

From now on, we shall assume that Ω has more than one component.

We treat first the special case $\Omega = \Omega_1 \cup \Omega_2$, with Ω_1, Ω_2 disjoint closed discs.
Subject both $\partial\Omega_1$, $\partial\Omega_2$ to the geodesic curvature flow to obtain families of discs
$(\Omega_1)_t$, $(\Omega_2)_t$. Let t_1 and t_2 be the times for Ω_1, Ω_2, respectively, to deform to
points.

We assume first the case $\Omega_1 \cap (\Omega_2)_t = \emptyset$ for all $t \in [0, t_2)$. Consider the family

$$\Omega_t = \begin{cases} \Omega_1 \cup (\Omega_2)_t, & 0 \leqslant t \leqslant t_2, \\ (\Omega_1)_{t-t_2}, & t_2 \leqslant t \leqslant t_1 + t_2. \end{cases}$$

Let $A(t) = A(\Omega_t)$. Since $A'(t) < 0$, we consider the inverse function $t(a)$. Then

$$\frac{dL^2(\partial\Omega_{t(a)})}{da} \geqslant 2\left(2\pi \#\{\text{components of } \Omega_{t(a)}\} - \int_{\Omega_{t(a)}} K\,dM\right)$$

$$\geqslant 2\left(2\pi - \int_{\Omega_{t(a)}} K\,dM\right)$$

$$= \frac{dL^2(B_{r(a)})}{da}.$$

Observe that the inequality in the second line is strict for small t (for a close to $A(\Omega)$). Hence integrating we have

$$L(\partial\Omega) = L(\partial\Omega_1) + L(\partial\Omega_2) > L(B_{r(A(\Omega))}).$$

So in this case Ω cannot be an isoperimetric region.

As in the previous case, consider two disjoint discs Ω_1 and Ω_2, but now assume that there is $t_0 < t_2$ so that $\Omega_1 \cap (\Omega_2)_{t_0} \neq \emptyset$. We consider some $x_0 \in \partial\Omega_1 \cap (\Omega_2)_{t_0}$. Let α_1, β be parametrizations of $\partial\Omega_1$, $\partial(\Omega_2)_{t_0}$, respectively, so that $\alpha_1(0) = \beta(0) = x_0$ and $\alpha'(0) = \beta'(0)$. For s small enough, consider the geodesic segments $\sigma_1 = \overline{\alpha(s)\beta(s)}$, and $\sigma_2 = \overline{\alpha(-s)\beta(-s)}$. By the triangle inequality,

$$L(\sigma_1) + L(\sigma_2) < L(\alpha|_{[-s,s]}) + L(\beta|_{[-s,s]}).$$

Hence the closed curve γ obtained from α, β, σ_1, σ_2 by removing the intervals $\alpha|_{[-s,s]}$, $\beta|_{[-s,s]}$ is a closed curve with length strictly less than $L(\partial\Omega_1)+L(\partial(\Omega_2)_{t_0})$ and area less than or equal to $A(\Omega_1) + A((\Omega_2)_{t_0})$. Round off the corners of γ to obtain a closed smooth curve enclosing a disc $\Omega^{(1)}$ with $L(\partial\Omega^{(1)}) < L(\partial\Omega_1) + L(\partial(\Omega_2)_{t_0})$, and $A(\Omega^{(1)}) > A(\Omega_1) + A((\Omega_2)_{t_0})$. Let $t_0 - \varepsilon$ be the time for which $A(\Omega^{(1)}) = A(\Omega_1) + A((\Omega_2)_{t_0-\varepsilon})$. We have $L(\partial\Omega^{(1)}) < L(\partial\Omega_1) + L(\partial(\Omega_2)_{t_0-\varepsilon})$. If $t^{(1)}$ is the time for $\Omega^{(1)}$ to collapse to a point, we define the family of domains

$$(\Omega)_t = \begin{cases} \Omega_1 \cup (\Omega_2)_t, & 0 \leqslant t \leqslant t_0 - \varepsilon, \\ (\Omega^{(1)})_{t-(t_0-\varepsilon)}, & t_0 - \varepsilon < t \leqslant t_0 - \varepsilon + t^{(1)}. \end{cases}$$

The area of $(\Omega)_t$ is decreasing and continuous and the length of $\partial(\Omega)_t$ is decreasing with a discontinuity at $t = t_0 - \varepsilon$. As in the previous case, we get

$$L(\partial\Omega) > L(\partial B_{r(A(\Omega))}).$$

Assume now that $\Omega = \bigcup_{i=1}^{k} \Omega_i$, $k \geqslant 2$, with Ω_i discs. Apply the geodesic curvature flow to $\partial\Omega_k$ until it either collapses to a point or touches for the first time t_0^k another connected component, which we may assume is Ω_{k-1}. Then replace $(\Omega_k)_{t_0^k} \cap \Omega_{k-1}$ by a smooth disc as in the previous paragraph, and apply the geodesic curvature flow to this new disc. By this procedure, we decrease the area continuously. Finally Ω collapses to a point and we may conclude that

$$L(\partial\Omega) > L(\partial B_{r(A(\Omega))}),$$

as in the previous case.

Finally, assume that $\Omega = \bigcup_{i=1}^{k} \Omega_i$, $k \geqslant 2$, so $\Omega_1, \ldots, \Omega_m$, with $1 \leqslant m \leqslant k$, are non-simply connected components of Ω. Each Ω_i, $i \leqslant m$, is a disc D_i from which some smooth discs have been removed, and

$$L(\partial\Omega_i) > L(\partial D_i), \qquad A(\Omega_i) < A(D_i), \qquad i \geqslant m.$$

Apply the deformation discussed in the previous paragraph to $\Omega' = \bigcup_{i=1}^{m} D_i \cup \bigcup_{i=m+1}^{k} \Omega_i$ observing that some Ω_i, for $i \geqslant m$, perhaps could have disappeared in this process (as some Ω_i could be included in some D_j). In this way we construct a deformation Ω'_t of continuously decreasing area and decreasing boundary length. Let t_0 be so that $A(\Omega'_{t_0}) = A(\Omega)$. Then we have

$$L(\partial\Omega) > L(\Omega') \geqslant L((\Omega')_{t_0}) \geqslant L(B_{r(A(\Omega))}),$$

and the proof is complete.   □

The proof included here corresponds to Theorems 2, 4, and 5 in the paper by Benjamini and Cao [15]. They also obtained a Riemannian comparison theorem for isoperimetric profiles.

Theorem 2.3.2 ([15, Theorem 8]). *Let M be a rotationally symmetric surface with Riemannian metric $g = dr^2 + f(r)^2 d\theta^2$ such that $\int_M K^+ dM \leqslant 2\pi$ and $K(r)$ is a decreasing function of r. Suppose that N is a simply connected Riemannian surface which is convex at infinity and that N satisfies*

$$\sup_{A(\Omega)=A(B_r)} \int_\Omega K_N \, dN \leqslant \int_{B_r} K_M \, dM$$

for each r. Then $I_N(a) \geqslant I_M(a)$ for each a. Moreover, if there exists a domain Ω with the same area as B_r and $L(\partial\Omega) = L(\partial B_r)$, then Ω must be isometric to B_r, where B_r is the geodesic disc of radius r centered at the origin in M.

Proof. In case $\int_{B_r} K \, dM < 2\pi$ for every $r > 0$, the proof of Theorem 2.3.1 applies without changes. So assume that $\int_B K \, dM = 2\pi$ for some rotationally symmetric ball B. Let $r_0 = \inf\{r : \int_{B_r} K \, dM = 2\pi\}$ and $r_1 = \sup\{r : \int_{B_r} K \, dM = 2\pi\}$. Since K is decreasing we have that $K \equiv 0$ in $\overline{B}_{r_1} - B_{r_0}$. On the other hand, for $r \in [r_0, r_1]$, we have

$$2\pi f'(r) = \int_{\partial B_r} h = 2\pi - \int_{B_r} K \, dM = 0.$$

Hence we conclude that $\overline{B}_{r_1} - B_{r_0}$ is isometric to a flat cylinder that is foliated by the geodesics $\{\partial B_r\}$, $r \in [r_0, r_1]$.

Let $C \subset M$ be a simple closed geodesic, and Ω the disc enclosed by C. By the Gauss–Bonnet Theorem, we have

$$0 = \int_C h = 2\pi - \int_\Omega K \, dM.$$

Since K is monotone, it follows from the definition of r_0 and r_1 that $C = \partial\Omega \subset \overline{B}_{r_1} - B_{r_0}$. By the uniqueness of geodesics (or the maximum principle) we have that $C = \partial B_r$ for some $r \in [r_0, r_1]$.

Let now $\Omega \subset N$. If $A(\Omega) < A(B_{r_0})$, then, by the assumption on the curvature of N, we have $\int_{\Omega} K_N \, dN \leqslant \int_{B_{r_0-\varepsilon}} K \, dM < 2\pi$, where $A(B_{r_0-\varepsilon}) = A(\Omega)$. The proof of the inequality $L(\partial\Omega) \geqslant L(\partial B_{r_0-\varepsilon})$ goes in the same way as in the proof of Theorem 2.3.1.

Assume now that $A(\Omega) \geqslant A(B_{r_0})$. To reproduce the arguments in the proof of Theorem 2.3.1 we have to take into account the case in which the geodesic curvature flow takes a Jordan curve to a simple closed geodesic. So assume that $\Omega \subset N$ is a disc. Apply the geodesic curvature flow to $\partial\Omega$ and suppose that there is convergence to a simple closed geodesic C. Let D be the disc enclosed by C. For $\lambda \geqslant 0$, let C_λ be the inner parallel to C at distance λ, and D_λ the enclosed disc. Observe that

$$\frac{d}{d\lambda} A(D_\lambda) = -L(C_\lambda),$$

and

$$\frac{d}{d\lambda} L(C_\lambda) = -\int_{C_\lambda} h_\lambda = -2\pi + \int_{D_\lambda} K \, dM \leqslant 0.$$

Observe also that if $A(D_\lambda) < A(B_{r_0})$, then $\int_{D_\lambda} K_N \, dN < 2\pi$. Then there is λ_0 so that C_{λ_0} is an embedded curve in N with non-zero geodesic curvature. Geodesic curvature flow is then applied to C_{λ_0} to continue the evolution. In either case, there is a family $\{\Omega_a\}$, $a \in [0, A(\Omega)]$, so that:

1. $A(\Omega_a) = a$, $A(\Omega_{A(\Omega)}) = \Omega$, and Ω_0 is a point.

2. $L(\partial\Omega_a)$ is non-decreasing.

Under these conditions, the proof of Theorem 2.3.1 can be used to prove the desired result. □

Benjamini and Cao also gave some applications of these results to get bounds for the Laplacian on domains in rotationally symmetric surfaces with non-increasing curvature [15, Thm. 10], and comparison results for eigenvalues, [15, Corollary B]. They also tried to apply their methods to rotationally symmetric spheres but, unfortunately, as stated in [70, p. 4904], their proof is wrong since equation (4.6) in [15] is wrong.

The methods employed in [15] were also used by Topping [117] to get similar results. Shortly after Benjamini and Cao's paper, Pansu [76] showed that the classical flow by parallels can be used to obtain the results in Benjamini and Cao's paper. Topping also recovered these results by polyhedral approximation [118]. Howards, Hutchings and Morgan [70] solved the isoperimetric problem in rotationally symmetric spheres in some particular cases. This problem was finally solved by the author [90] from a classification of simple closed curves with constant geodesic curvature, an approach used by Schmidt [104]. Using these techniques, Cañete [23] classified the stable regions in tori of revolution, and Cañete and the author [20] completed the classification of isoperimetric regions in quadrics of revolution. A generalization of the Banchoff–Pohl inequality using the curve shortening flow was given by Süssmann [116].

2.3.2   Spheres

As remarked earlier, it is hard to obtain sharp isoperimetric inequalities in spheres with Gauss curvature decreasing from the poles by using the curve shortening flow. In this subsection, following Morgan and Johnson [71], we give a slight improvement of the Bol–Fiala inequality in spheres with Gauss curvature bounded from above by a positive constant K_0. The inequality involves the classical Bol–Fiala inequality and the length of the shortest closed geodesic. We begin with a preliminary result

Lemma 2.3.3. *Let M be a smooth Riemannian surface with $K \leqslant 1$. Let C be a smooth Jordan curve of length L which flows under the curve shortening flow to a point and encloses area A. If $L^2 \leqslant 4\pi$, then*

$$A \leqslant 2\pi - (4\pi^2 - L^2)^{1/2}.$$

Proof. Whenever $dA/dt \neq 0$ we have

$$\frac{dL^2}{dA} \geqslant 4\pi - 2A.$$

Since L is decreasing along the flow, we may integrate this differential inequality with respect to A to get $A^2 - 4\pi A + L^2 \geqslant 0$ along the deformation. The two roots of this polynomial in A are $2\pi \pm (4\pi^2 - L^2)^{1/2}$. Hence $A \geqslant 0$ if and only if

$$\text{either } A \leqslant 2\pi - (4\pi^2 - L^2)^{1/2}, \text{ or } A \geqslant 2\pi + (4\pi^2 - L^2)^{1/2}.$$

As A approaches 0 when the flowing curves converge to the limit point, we find that the first equality is the one that really holds.   □

The following lemma will also be used in the proof of the main result of this section.

Lemma 2.3.4. *Let $f\colon [0, r] \to \mathbb{R}$ be a concave function. Assume that $a_1, \ldots, a_k$ and $a_1 + \cdots + a_k$ are in $[0, r]$. Then we have*

$$\sum_{i=1}^{k} \left(f(0) - f(a_i) \right) \leqslant f(0) - f\left(\sum_{i=1}^{k} a_i \right).$$

Proof. It is a repeated application of the inequality $f(0) - f(a) \leqslant f(b) - f(a+b)$, which is obtained from

$$\frac{f(b) - f(0)}{b} \geqslant \frac{f(a+b) - f(a)}{b}.$$
□

Theorem 2.3.5 ([71, Thm. 5.3]). *Let M be a smooth Riemannian sphere with Gauss curvature $K \leqslant K_0$. Let L_0 be the infimum of the length of simple closed geodesics. Let A, L be the area and perimeter of a region Ω with smooth boundary C. Then*

$$L^2 \geqslant \min\{(2L_0)^2, 4\pi A - K_0 A^2\}.$$

Proof. By scaling the Riemannian metric on M we may assume that $K_0 = 1$. Let $C_1, \ldots, C_k$ denote the connected components of C. Let us assume that $L^2 < \max\{4\pi A - A^2\} = 4\pi$ and also that $L < 2L_0$. The last condition implies that, with the possible exception of C_1, all the components of C flow to points. By Lemma 2.3.3, each curve C_i, $i \geqslant 2$, bounds a disc D_i satisfying

$$A_i < 2\pi - (4\pi^2 - L_i^2)^{1/2},$$

where L_i is the length of ∂D_i and A_i the area of D_i.

From the Bol–Fiala inequality (2.2)

$$L_1^2 \geqslant 4\pi A_1 - A_1^2,$$

and we obtain $(2\pi - A_1)^2 \geqslant 4\pi^2 - L_1^2$. Hence one of the following inequalities holds:

$$A_1 \leqslant 2\pi - (4\pi^2 - L_1^2)^{1/2},$$
$$A_1 \geqslant 2\pi + (4\pi^2 - L_1^2)^{1/2}.$$

Assume A_1 satisfies $A_1 \leqslant 2\pi - (4\pi^2 - L_1^2)^{1/2}$. Then

$$A \leqslant \sum_{i=1}^{k} A_i \leqslant \sum_{i=1}^{k} 2\pi - (4\pi^2 - L_i^2)^{1/2} \leqslant 2\pi - (4\pi^2 - L^2)^{1/2},$$

by Lemma 2.3.4, since $x \mapsto (4\pi^2 - x^2)^{1/2}$ is concave. This implies the desired inequality $L^2 \geqslant 4\pi A - A^2$.

If C_1 satisfies $A_1 \geqslant 2\pi + (4\pi^2 - L_1^2)^{1/2}$, then $D_i \subset D_1$ for all $i \geqslant 2$ by the avoidance principle. Then

$$A \geqslant A_1 - \sum_{i=2}^{k} A_i \geqslant \left(2\pi + (4\pi^2 - L_1^2)^{1/2}\right) - \left(2\pi - (4\pi^2 - Q^2)^{1/2}\right)$$

$$\geqslant (4\pi^2)^{1/2} + (4\pi^2 - (L_1 + Q)^2)^{1/2} = 2\pi + (4\pi^2 - L^2)^{1/2},$$

by two more applications of Lemma 2.3.4. This implies again that the inequality $L^2 \geqslant 4\pi A - A^2$ holds. $\square$

Chapter 3

Higher dimensions

3.1 Introduction

Unlike surfaces, the use of the mean curvature flow in higher dimensions to prove isoperimetric inequalities is severely limited by the possibility of development of singularities. The reader is referred to Sinestrari's course in this volume [113] for an updated discussion on these topics.

On the other hand, isoperimetric inequalities can be used to ensure the singularity formation for given initial data. A typical argument is the one given by Topping [117, §4]. Given an immersed surface $\Sigma \subset \mathbb{R}^3$, we define its *Willmore energy* by

$$\int_\Sigma H^2 \, d\Sigma.$$

It is known that, for any immersed surface $\Sigma \subset \mathbb{R}^3$, the Willmore energy satisfies $\int_\Sigma H^2 \, d\Sigma \geqslant 4\pi$ and that equality holds if and only if Σ is a round sphere, since one can estimate H^2 from below by K^+, and the integral of K^+ is larger than or equal to 4π; see the introduction of [27] and [63]. Simon [111] proved that there exists a torus minimizing the Willmore energy over all tori. Let θ_0 be the minimum of the Willmore energy in the class of immersed tori in $\mathbb{R}^3$. We obviously have $\theta_0 > 4\pi$.

Assume that $\Sigma \subset \mathbb{R}^3$ is an embedded torus evolving by mean curvature without developing singularities and that the volume enclosed by Σ goes to 0 when we approach the extinction time T. Calling $A(t)$ the area of Σ_t and $V(t)$ the volume enclosed by Σ_t, we have

$$-\frac{dV}{dt} = \int_{\Sigma_t} H_t \, d\Sigma_t \leqslant \left(\int_{\Sigma_t} H_t^2 \, d\Sigma_t \right)^2 A^{1/2} \leqslant \frac{A^{1/2}}{\theta_0^{1/2}} \int_{\Sigma_t} H_t^2 \, d\Sigma_t,$$

$$-\frac{dA^{3/2}}{dt} = 3A^{1/2} \int_{\Sigma_t} H_t^2 \, d\Sigma_t.$$

This implies

$$\frac{dA^{3/2}}{dt} - 3\theta_0^{1/2}\frac{dV}{dt} \leqslant 0.$$

Integrating with respect to t from 0 to T we get

$$A(\Sigma)^{3/2} - 3\theta_0^{1/2}V(\Omega) \geqslant \lim_{t\to T}\left\{A(\Sigma_t)^{3/2} - 3\theta_0^{1/2}V(\Omega_t)\right\} = \lim_{t\to T}A(\Sigma_t)^{3/2} \geqslant 0.$$

As $\theta_0 > 4\pi$, we obtain

$$A(\Sigma)^3 > 9\theta_0\,V(\Omega)^2 > 36\pi\,V(\Omega)^2.$$

If we pick a torus $\Sigma \subset \mathbb{R}^3$ so that $A(\Sigma)^3$ is very close to and smaller than $9\theta_0\,V(\Omega)^2$ (as a round sphere with a very thin tube attached), then the mean curvature flow applied to Σ will certainly develop singularities.

Figure 3.1: A torus close to a round sphere satisfying $A(\Sigma)^3 < 9\theta_0\,V(\Omega)^2$. By the previous arguments, it will develop singularities when evolving under the mean curvature flow

3.2 H^k-flows and isoperimetric inequalities

Recently, F. Schulze [106] has been able to produce a new proof of the isoperimetric inequality in Euclidean space (up to dimension 8), and also to recover Kleiner's result on three-dimensional Hadamard manifolds, using a level-set formulation of (a power of) the mean curvature flow. In his work, a major role is played by a generalization of the inequality $\int_\Sigma H^2 d\Sigma \geqslant 4\pi$ to higher dimensions and to boundaries of sets with low regularity.

The essential idea behind Schultze's result is the following. Given a compact Riemannian manifold Σ^n without boundary, an N^{n+1} complete Riemannian manifold, and an embedding $F_0 \colon \Sigma \to N$, the H^k-flow or flow by the k-th power of the mean curvature is given by a smooth function $F \colon \Sigma \times [0,T) \to N$ such that

$$\begin{cases} F(\cdot,0) = F_0(\cdot), \\ \dfrac{dF}{dt}(\cdot,t) = H^k(\cdot,t)\, N(\cdot,t), \end{cases} \qquad (3.1)$$

where $k \geqslant 1$, $H(\cdot,t)$ is the mean curvature of the immersion $F_t(\cdot) = F(\cdot,t)$, and $N(\cdot,t)$ is the inner unit normal to F_t. An interesting property of the flow (3.1) is the following.

Proposition 3.2.1 ([106, §2]). *Along a solution of the flow by the k-th power of the mean curvature, the isoperimetric difference*

$$A(t)^{(n+1)/n} - c_{n+1}\, V(t)$$

is non-increasing if $k + 1 \geqslant n$ and the inequality

$$\int_{\Sigma_t} |H|^n \geqslant A(\mathbb{S}^n) \qquad (3.2)$$

holds. Here $A(t) = \mathcal{H}^n(F_t(\Sigma))$, $V(t)$ is the $\mathcal{H}^{n+1}$-measure of the region enclosed by $\Sigma_t = F_t(\Sigma)$, and H_t is the mean curvature of $F_t \colon \Sigma \to N$.

Let us check the monotonicity of the isoperimetric difference when the flow does not develop singularities. So consider an immersed hypersurface $\Sigma \subset \mathbb{R}^{n+1}$, and let it evolve by the H^k curvature flow, where $k \geqslant n-1$ is an integer. We also assume that Σ is mean convex $(H > 0)$, and that Σ collapses to zero volume in finite time. Let us compute the derivative of the isoperimetric difference

$$A^{(n+1)/n} - c_{n+1}\, V$$

along the flow. Here c_{n+1} is the Euclidean isoperimetric constant

$$c_{n+1} = \frac{c_n^{(n+1)/n}}{\omega_{n+1}} = \frac{A(\mathbb{S}^n)^{(n+1)/n}}{V(\mathbb{B}^{n+1})}.$$

We have

$$-\frac{dA^{(n+1)/n}}{dt} = \frac{n+1}{n}\, A^{1/n} \int_{\Sigma_t} (-nH_t)\, H_t^k \, d\Sigma_t = -(n+1)\, A^{1/n} \int_{\Sigma_t} H_t^{k+1}\, d\Sigma_t,$$

and

$$-\frac{dV}{dt} = \int_{\Sigma_t} H_t^k\, d\Sigma_t \leqslant \left(\int_{\Sigma_t} H_t^{k+1}\, d\Sigma_t \right)^{k/(k+1)} A^{1-k/(k+1)}.$$

Now we use the *Willmore inequality* for immersed hypersurfaces in $\mathbb{R}^{n+1}$, proved by Chen [27]. If $\Sigma \subset \mathbb{R}^{n+1}$ is a compact immersed hypersurface, then

$$\int_{\Sigma} |H|^n\, d\Sigma \geqslant A(\mathbb{S}^n), \qquad (3.3)$$

and equality holds if and only if Σ is a round sphere. From (3.3) and the Hölder inequality, as $n/(k+1) \leqslant 1$, we get

$$1 \leqslant A(\mathbb{S}^n)^{-1/n} \left(\int_{\Sigma_t} H_t^n \, d\Sigma_t \right)^{1/n} \leqslant A(\mathbb{S}^n)^{-1} \left(\int_{\Sigma_t} H_t^{k+1} \right)^{1/(k+1)} A^{1/n - 1/(k+1)}.$$

So we obtain

$$-\frac{dV}{dt} \leqslant \left(\int_{\Sigma_t} H_t^{k+1} \, d\Sigma_t \right) A^{1/n} A(\mathbb{S}^n)^{-1/n},$$

and hence

$$\frac{dA^{(n+1)/n}}{dt} - c_{n+1} \frac{dV}{dt} \leqslant A^{1/n} \left(\int_{\Sigma_t} H_t^{k+1} \, d\Sigma_t \right) ((n+1) - c_{n+1} A(\mathbb{S}^n)^{-1/n}) = 0,$$

since

$$c_{n+1} = \frac{A(\mathbb{S}^n)^{(n+1)/n}}{V(\mathbb{B}^{n+1})} = \frac{A(\mathbb{S}^n)^{(n+1)/n}}{A(\mathbb{S}^n)/(n+1)} = (n+1) \, A(\mathbb{S}^n)^{1/n}.$$

Integrating the derivative of the isoperimetric difference between 0 and T, since Σ collapses to zero volume, we have that the isoperimetric inequality holds for Σ.

Since the classical flow (3.1) may develop singularities, a level-set formulation, in which (3.1) is replaced by

$$\operatorname{div} \left(\frac{\nabla u}{|\nabla u|} \right) = -\frac{1}{|\nabla u|^{1/k}}, \tag{3.4}$$

is considered by Schultze. Here Ω is a bounded open set with $H|_{\partial \Omega} > 0$, $u \colon \overline{\Omega} \to \mathbb{R}$ and $u = 0$ in $\partial \Omega$. The evolving surfaces for this formulation are then given as level sets of u, i.e., by $\Gamma_t = \partial \{ x \in \Omega : u(x) > t \}$.

The main results in Schulze's paper are the following.

Theorem 3.2.2 ([106, Corollary 1.2]). *Let $\Omega \subset \mathbb{R}^{n+1}$ be a compact domain with smooth boundary and $n+1 \leqslant 8$, or $\Omega \subset N^3$, where N^3 is a simply connected complete manifold with non-negative sectional curvatures. Then*

$$\mathcal{H}^n(\partial \Omega)^{(n+1)/n} \geqslant c_{n+1} \mathcal{H}^{n+1}(\Omega),$$

where $\mathcal{H}^k$ is the k-dimensional Hausdorff measure.

The result for 3-dimensional Hadamard manifolds was proved by B. Kleiner [56]; see Section 3.3.2 for an alternative proof. It is a particular case of Aubin's conjecture, or the Cartan–Hadamard conjecture [9], that states that the Euclidean isoperimetric inequality holds in a *Hadamard manifold*, i.e., a complete, simply connected manifold with non-negative sectional curvatures [9], [46, §6.28$\frac{1}{2}$]. The 4-dimensional case of the conjecture was proved by Croke [33].

Schulze's proof of the isoperimetric inequality follows from the next result.

Theorem 3.2.3 ([106, Thm. 1.1]). *Let $\Omega \subset \mathbb{R}^n$ be a bounded open subset of N^{n+1} such that $H|_{\partial\Omega} > 0$. If $n = 2$, let $k \geqslant 1$ and N be a 3-dimensional Hadamard manifold. If $n \geqslant 3$, let $N = \mathbb{R}^{n+1}$ and $k > n$. If u is a weak H^k-flow generated by Ω, then the isoperimetric difference*

$$\left(\mathcal{H}^n(\partial^*\{u > t\})\right)^{(n+1)/n} - c_{n+1}\,\mathcal{H}^{n+1}(\{u > t\})$$

is a non-negative decreasing function on $[0, T)$, where $T = \sup_\Omega u$.

Equation (3.4) is degenerate elliptic, and can be solved by elliptic regularization considering the problems

$$\operatorname{div}\left(\frac{\nabla u^\varepsilon}{\sqrt{\varepsilon^2 + |\nabla u^\varepsilon|^2}}\right) = -(\varepsilon^2 + |\nabla u^\varepsilon|^2)^{-1/2k} \quad \text{in } \Omega,$$
$$u^\varepsilon = 0 \qquad\qquad\qquad\qquad \text{in } \partial\Omega. \tag{3.5}$$

Uniform L^∞ and gradient bounds are obtained in [106, §3] so that there is convergence of subsequences of $\{u^\varepsilon\}$ to a Lipschitz function $u\colon \overline{\Omega} \to \mathbb{R}$. This u is called a weak solution of (3.4). Uniqueness of this weak solution is proved in $\mathbb{R}^{n+1}$ for $n \leqslant 6$ [106, Corollary 4.5]. By the results in [106, §5], there is a set $B \subset [0, T]$ of full measure such that, for $t \in B$, up to a closed set of $\mathcal{H}^n$-measure 0, the set $\Gamma_t = \partial\{u > t\}$ is a $C^{1,\alpha}$ hypersurface which carries a weak mean curvature in $L^{k+1}(\Gamma_t)$. The main point is now to get the estimate (3.2) on these Γ_t. This is done by considering the parallel hypersurfaces. In Hadamard manifolds, (3.2) is obtained from a modification of an argument by Simon [111]. Finally, in [106, §7], the monotonicity of the isoperimetric difference is proved.

3.3 Estimates on the Willmore functional and isoperimetric inequalities

However, estimate (3.2) is enough to get the isoperimetric inequality in Euclidean space for any dimension, and for a 3-dimensional Hadamard manifold. Observe that (3.2) was obtained by [27] for smooth submanifolds of the Euclidean space, and by Almgren [4] for solutions of the Plateau problem. The key point was observed by Kleiner [56], who showed that estimates of the Willmore functional of a hypersurface $\Sigma^n \subset N^{n+1}$,

$$\int_\Sigma |H|^n d\Sigma,$$

provide bounds for the isoperimetric profile of N.

3.3.1 Euclidean spaces

Let us see how to get the Euclidean isoperimetric inequality from an estimate of this Willmore functional. Consider the problem of minimizing perimeter in $\mathbb{R}^{n+1}$

under a volume constraint. There is always a solution Ω enclosing a given volume v since the quotient of $\mathbb{R}^{n+1}$ by its isometry group is compact [66], [68]. The set Ω is bounded by the monotonicity formula [68]. It is known that the boundary Σ of Ω is smooth and has positive constant mean curvature except on a set $\Sigma_0 \subset \Sigma$ of Hausdorff dimension less than or equal to $n - 7$ [43]. At every singular point $p \in \Sigma_0$, there is a minimizing tangent cone [42], that cannot be contained in a halfspace of $\mathbb{R}^{n+1}$ [18]. For every unit vector $v \in \mathbb{S}^n$, there is a support hyperplane Π_v of Ω at some given point p_v, which cannot be singular (since the minimizing tangent cone would be contained in a halfspace). Let Σ^+ be the set of regular points with non-negative principal curvatures. Then $p_v \in \Sigma^+$, and we have

$$\int_\Sigma H^n \, d\Sigma \geqslant \int_{\Sigma^+} H^n \, d\Sigma \geqslant \int_{\Sigma^+} GK \, d\Sigma \geqslant A(\mathbb{S}^n), \tag{3.1}$$

where GK is the Gauss–Kronecker curvature (the product of the principal curvatures), and in the last inequality we have used the area formula and the fact that the Jacobian of the Gauss map is the Gauss–Kronecker curvature. Equality holds if and only is Σ is totally umbilical without singular points, i.e., if and only if it is a totally umbilical sphere.

Let us denote the isoperimetric profile of $\mathbb{R}^{n+1}$ by I. By standard arguments ([55], pp. 170–172), we have:

- I is continuous and increasing,

- if I is smooth at v_0, then $I'(v_0) = nH$, where H is the constant mean curvature of *any* isoperimetric region of volume v_0, and

- left and right derivatives of I exist everywhere.

Since I is a continuous monotone function with left and right derivatives at every point, it is absolutely continuous.

The continuity of I follows from the convergence of isoperimetric regions. To prove the monotonicity of I, we just need to show that the constant mean curvature H of the boundary of an isoperimetric region Ω is positive, which follows easily from the maximum principle for the constant mean curvature equation. To compute the derivative $I'(v_0)$, we simply make a variation supported in a neighborhood of a regular point of Σ and use the first variation formulas for volume and area. The fact that left and right derivatives always exist follows from the convergence of isoperimetric regions of volumes $v_k \to v_0$ to an isoperimetric region of volume v_0; see [55], p. 171.

Let $J(v)$ be the perimeter of a ball of volume v. Let $f(a)$, $g(a)$ be the inverse functions of I, J, respectively. We know that $g'(a) = J'(a)^{-1} = (2H(a))^{-1}$, where $H(a)$ is the mean curvature of a sphere of area a. When f' exists, $f'(a) = I'(a)^{-1} = (2H)^{-1}$, where H is the mean curvature of any isoperimetric region of volume $f(a)$. By inequality (3.1) and the characterization of equality, we have $g'(a) \geqslant f'(a)$ $\mathcal{L}^1$-a.e. Since f is absolutely continuous (and g is smooth), we have $g(a) \geqslant f(a)$. It follows that $I \geqslant J$.

If equality holds for some v_0, then for $a_0 = J(v_0) = I(v_0)$ we have $g(a_0) = f(a_0)$. Since $g' \geqslant f'$, we obtain that $f \equiv g$ in the interval $(0, a_0)$, and so the mean curvature H of any isoperimetric region Ω_0 of area a_0 satisfies $H^{-1} = H(a_0)^{-1}$. If Ω_0 is any isoperimetric region of volume v_0, then Ω_0 is isometric to a ball in $\mathbb{R}^{n+1}$ of volume v_0.

3.3.2 3-dimensional Hadamard manifolds

Let us consider now the case of a 3-dimensional Hadamard manifold M^3. The arguments in the following are taken from [92]. We have:

Theorem 3.3.1 ([92]). *Let $\Sigma \subset M$ be a compact $C^{1,1}$ surface embedded in a 3-dim-ensional Hadamard manifold M. Then*

$$\int_\Sigma H^2 \, d\Sigma \geqslant 4\pi. \tag{3.2}$$

If equality holds in (3.2), then Σ bounds a flat region. Moreover, (3.2) implies that

$$\left(\max_\Sigma H^2 \right) A(\Sigma) \geqslant 4\pi. \tag{3.3}$$

Equality holds in (3.3) if and only if Σ bounds a region isometric to a Euclidean ball. If $K_{\sec} \leqslant -1$, then

$$\int_\Sigma (-1 + H^2) \, d\Sigma \geqslant 4\pi. \tag{3.4}$$

If equality holds in (3.4), then Σ bounds a hyperbolic region. Moreover, from (3.4) we obtain

$$\left(\max_\Sigma (-1 + H^2) \right) A(\Sigma) \geqslant 4\pi. \tag{3.5}$$

Equality holds in (3.5) if and only if Σ bounds a region isometric to a hyperbolic ball of sectional curvatures equal to -1.

Proof. Consider first the case $K_{\sec} \leqslant 0$. Let g be the Riemannian metric on M. Let Σ be an embedded $C^{1,1}$ compact surface. Take $p \in \Sigma$ and let $d(q)$ denote the Riemannian distance between q and p. For $\varepsilon > 0$, we consider the family of conformal metrics

$$g_\varepsilon = \rho_\varepsilon^2 g = e^{2u_\varepsilon} g,$$

where

$$\rho_\varepsilon = \frac{2\varepsilon}{1 + \varepsilon^2 d^2}, \qquad u_\varepsilon = \log \left(\frac{2\varepsilon}{1 + \varepsilon^2 d^2} \right).$$

In case M is the Euclidean space, this metric is obtained by applying a conformal transformation to the metric of the sphere and projecting this metric orthogonally to the Euclidean space by means of stereographic projection. It was used by Li and

Yau [63] to obtain lower bounds for the Willmore functional in Euclidean 3-space. By taking into account the well-known relation between conformal metrics, we get

$$e^{2u_\varepsilon}(K_\varepsilon)_{\text{sec}} \geqslant K_{\text{sec}} + e^{2u_\varepsilon}, \tag{3.6}$$

where K_ε and K are the sectional curvatures of a given plane of M for the metrics g_ε and g, respectively. From now on we shall assume that they are the ones of the tangent plane to Σ. Hence,

$$
\begin{aligned}
\int_\Sigma H^2 \, d\Sigma &= \int_\Sigma (H^2 + K_{\text{sec}}) \, d\Sigma - \int_\Sigma K_{\text{sec}} \, d\Sigma \\
&= \int_\Sigma ((H_\varepsilon^2) + (K_\varepsilon)_{\text{sec}}) \, d\Sigma_\varepsilon - \int_\Sigma K_{\text{sec}} \, d\Sigma \\
&\geqslant \int_\Sigma H_\varepsilon^2 \, d\Sigma_\varepsilon + \int_\Sigma d\Sigma_\varepsilon \\
&\geqslant \int_\Sigma d\Sigma_\varepsilon,
\end{aligned}
$$

where in the first equality we use the conformal invariance of $\int(H^2 + K_{\text{sec}}) \, d\Sigma$ and inequality (3.6). The area element of Σ with respect to the metric g_ε is denoted by $d\Sigma_\varepsilon$.

The limit

$$\lim_{\varepsilon \to 0} \int_\Sigma d\Sigma_\varepsilon$$

can be computed in (ambient) polar coordinates. Let $r(q) = d(q)$. Fix $r_0 > 0$ small. Observe that e^{2u_ε} converges uniformly to 0 in Σ out of the ball $B(p, r_0)$ when $\varepsilon \to \infty$. For r_0 small enough, the modulus of the gradient of $r|_\Sigma$ over $\Sigma \cap B(p, r_0)$ is approximately 1, and the length of $\Sigma \cap \partial B(p, r)$, for $r \in (0, r_0)$, is approximately $2\pi r$. By applying the coarea formula to $r|_\Sigma$ we get

$$\int_{\Sigma \cap B(p, r_0)} d\Sigma_\varepsilon = 2\pi \int_0^{r_0} r \left(\frac{2\varepsilon}{1 + \varepsilon^2 r^2} \right)^2 dr + o(r_0) = 2\pi + o(r_0),$$

where $o(r_0)$ converges to 0 when $r_0 \to 0$. Letting $\varepsilon \to \infty$ and $r_0 \to 0$ we get

$$\lim_{\varepsilon \to \infty} \int_\Sigma d\Sigma_\varepsilon = 4\pi,$$

and we obtain (3.2).

When equality holds in (3.2) we need a more accurate estimate of the expression of the curvatures in the conformal metrics. So we write

$$
\begin{aligned}
e^{2u_\varepsilon} K_\varepsilon = K &- \left(\frac{\varepsilon^2}{1 + \varepsilon^2 d^2} \right)^2 4d^2 \\
&+ \left(\frac{\varepsilon^2}{1 + \varepsilon^2 d^2} \right) (\nabla^2 d^2(X, X) + \nabla^2 d^2(Y, Y)),
\end{aligned}
$$

where X, Y is an orthonormal basis of the tangent plane to Σ. From this formula we get

$$
\begin{aligned}
4\pi &= \int_\Sigma H^2 \, d\Sigma = \int_\Sigma (H^2 + K) \, d\Sigma - \int_\Sigma K \, d\Sigma \\
&= \int_\Sigma ((H_\varepsilon^2) + K_\varepsilon) \, d\Sigma_\varepsilon - \int_\Sigma K \, d\Sigma \\
&= \int_\Sigma d\Sigma_\varepsilon + \int_\Sigma \left(\frac{\varepsilon^2}{1 + \varepsilon^2 d^2} \right) (\nabla^2 d^2(X,X) + \nabla^2 d^2(Y,Y) - 4) \, d\Sigma \\
&\quad + \int_\Sigma H_\varepsilon^2 \, d\Sigma_\varepsilon.
\end{aligned}
$$

We already know that the first integral converges to 4π when $\varepsilon \to \infty$. So the limit of the remaining integrals is 0. Since $\nabla^2 d^2(X,X) \geqslant 2$ for any $|X| = 1$, we obtain that both integrals are positive and, in particular,

$$
\nabla^2 d^2(X,X) = \nabla^2 d^2(Y,Y) = 2.
$$

Standard comparison theorems in Riemannian geometry [81] show that, if the geodesic starting from p leaves the enclosed domain Ω in a non-tangential way, then $\nabla^2 d^2 = 2g$ at the hitting point. Standard comparison shows that $\nabla d^2 \equiv 2g$ along the geodesic. Moving slightly the geodesic we get a cone with the property that $\nabla d^2 \equiv 2g$ inside this cone. Since every point in the interior of Ω can be connected with Σ by a minimizing geodesic hitting Σ orthogonally, we conclude that every point inside Σ is flat and so Ω is flat.

In case equality

$$
\left(\max_\Sigma H^2 \right) A(\Sigma) = 4\pi
$$

holds, in addition to Ω flat we get that the mean curvature of Σ is constant. For any domain of this type, Ros [97] and Montiel and Ros [65] have proved that

$$
3\, V(\Omega) \leqslant \frac{1}{H} A(\Sigma),
$$

and equality holds if and only if Ω is isometric to a geodesic ball in Euclidean space. But the classical Minkowski formula

$$
3\, V(\Omega) = \frac{1}{H} A(\Sigma)
$$

holds in Ω, since the function $(1/2)\, d^2$ has Hessian on Ω proportional to twice the identity matrix. From this we conclude our proof of Theorem 3.3.1 in the flat case.

In the hyperbolic case $K_{\sec} \leqslant -1$, one has to consider the following family of conformal metrics:

$$
g_\varepsilon = \left(\frac{2\varepsilon}{(1 - \varepsilon^2) + (1 + \varepsilon^2)\cosh(d)} \right) g, \qquad \varepsilon > 1.
$$

In case M is the hyperbolic 3-space, they are obtained by writing the spherical metric in a disc D of $\mathbb{R}^3$ via stereographical projection in terms of the hyperbolic metric of constant curvature -1 in D. So we obtain

$$e^{2u_\varepsilon}(K_\varepsilon)_{\text{sec}} \geqslant K_{\text{sec}} + e^{2u_\varepsilon} + 1$$

and

$$\int_\Sigma (-1 + H^2)\, d\Sigma \geqslant \int_\Sigma d\Sigma_\varepsilon.$$

As in the previous case, one proves that

$$\lim_{\varepsilon \to \infty} \int_\Sigma d\Sigma_\varepsilon = 4\pi,$$

which yields (3.4).

To analyze the equality case in (3.4), it is more convenient to write

$$e^{2u_\varepsilon}(K_\varepsilon)_{\text{sec}} = K_{\text{sec}} + 1 + e^{2u_\varepsilon} + \left(\frac{1 + \varepsilon^2}{(1 - \varepsilon^2) + (1 + \varepsilon^2)\,\cosh(d)} \right)$$
$$\times \left(\nabla^2 \cosh(d)(X, X) + \nabla^2 \cosh(d)(Y, Y) - 2\cosh(d) \right).$$

We recall that, by classical comparison theorems [81], when $K_{\text{sec}} \leqslant -1$ we get

$$\nabla^2 \cosh(d) \geqslant \cosh(d)\, g,$$

so that the factor in the previous displayed line is non-negative. Hence

$$4\pi = \int_\Sigma (-1 + H^2)\, d\Sigma = \int_\Sigma d\Sigma_\varepsilon + \int_\Sigma H_\varepsilon^2\, d\Sigma_\varepsilon$$
$$+ \int_\Sigma \left(\frac{1 + \varepsilon^2}{(1 - \varepsilon^2) + (1 + \varepsilon^2)\,\cosh(d)} \right)$$
$$\times \left(\nabla^2 \cosh(d)(X, X) + \nabla^2 \cosh(d)(Y, Y) - 2\cosh(d) \right) d\Sigma.$$

Letting $\varepsilon \to \infty$ and taking into account that $\lim_{\varepsilon \to \infty} \int_\Sigma dA_\varepsilon = 4\pi$, we deduce that the remaining positive integrals tend to 0 when $\varepsilon \to \infty$. In particular,

$$\nabla^2 \cosh(d)(X, X) = \nabla^2 \cosh(d)(Y, Y) = \cosh(d).$$

By standard comparison theorems, and arguing as in the Euclidean case, we conclude that the metric in Ω is hyperbolic. If

$$\left(\max_\Sigma (-1 + H^2) \right) A(\Sigma) = 4\pi,$$

then H is constant. Moreover, from [64, Theorem 9] we conclude, by taking inner parallels, that

$$\int_\Sigma \left(\cosh(d) + H \, \sinh(d) \, \langle \partial/\partial d, N \rangle \right) d\Sigma \geqslant 0,$$

and equality holds only when Σ is a geodesic sphere. But, since the metric in Ω is hyperbolic, we have $\nabla^2 \cosh(d) = 2g$, so that

$$\int_\Sigma \left(\cosh(d) + H \, \sinh(d) \, \langle \partial/\partial d, N \rangle \right) d\Sigma = 0,$$

and Theorem 3.3.1 also follows in the hyperbolic case.   □

Once Theorem 3.3.1 is proven, the isoperimetric comparison theorem can be obtained following the arguments of the Euclidean case. However, a major difficulty is the non-existence of isoperimetric regions in non-compact manifolds. To overcome this, we use an argument by Kleiner [56] and take a fixed $p \in M$, and a sequence B_k of balls centered at p whose radii go off to infinity as $k \to \infty$. We consider the isoperimetric profile I_k of B_k, and J the isoperimetric profile of either the Euclidean space or hyperbolic space of constant sectional curvature -1. To these balls one can apply the following regularity result for isoperimetric regions [114].

Proposition 3.3.2. *Let B be a compact manifold with smooth boundary ∂B in a 3-dimensional manifold. Let $v \in (0, V(B))$. Then there is a region $\Omega \subset B$ with boundary $\Sigma = \partial \Omega$ such that*

1. *$V(\Omega) = v$, $A(\Sigma) = I_B(v)$.*

2. *$\Sigma = \partial \Omega$ is a $C^{1,1}$ surface in a neighborhood of ∂B.*

3. *Σ is C^∞ surface in the interior of Σ with constant mean curvature H.*

4. *The mean curvature h of Σ is defined almost everywhere (except in a set of $\mathcal{H}^2$-measure zero), and we have $h \leqslant H$.*

Hence there is an isoperimetric comparison theorem and we get $I_k(v) \geqslant J(v)$ for all $v \in (0, V(B_k))$. So we get $I_k(v) \geqslant J(v)$.

These ideas have been used by Choe and Ritoré [30], and Choe, Ghomi and Ritoré [28], [29] to prove optimal isoperimetric inequalities outside convex sets in 3-dimensional Hadamard manifolds, and Euclidean spaces, respectively. Comparison theorems for Ricci curvature bounded below have been given by Bayle [12] and Morgan and Johnson [71], and inside convex sets in manifolds with Ricci curvature bounded below by Bayle and Rosales [13] and Morgan [69].

3.4   Singularities in the volume-preserving mean curvature flow

Given a compact Riemannian manifold M^n without boundary, an N^{n+1} complete Riemannian manifold, and an embedding $F_0\colon M \to N$, the volume-preserving mean curvature flow is given by a smooth function $F\colon [0,T) \times M \to N$ so that

$$\begin{cases} F(\,\cdot\,,0) = F_0(\,\cdot\,), \\ \dfrac{dF}{dt}(\,\cdot\,,t) = (h - H)\,\nu(\,\cdot\,,t), \end{cases} \tag{3.1}$$

where

$$h(t) = \frac{\int_{M_t} H_t\, dM_t}{A(M_t)},$$

and H_t is the mean curvature (the trace of the Weingarten operator) of the immersion $F_t(\,\cdot\,) = F(\,\cdot\,,t)$, and $\nu(\,\cdot\,,t)$ is the inner unit normal to F_t. Volume-preserving mean curvature flow is of a global nature because of the term involving the total mean curvature of $M_t = F(M \times \{t\})$. Volume-preserving mean curvature flow can develop singularities in finite time. The classification of solutions for given isoperimetric problems can help in continuing the flow across the singularities. We briefly recall the results by Cabezas–Rivas and Miquel [21], [22] (see also Athanassenas [7], [8]). For simplicity we state their results in Euclidean space.

Theorem 3.4.1 ([22]). *Let $M \subset \mathbb{R}^{n+1}$ be a smoothly embedded hypersurface contained in the slab $G = \{x \in \mathbb{R}^{n+1} : 0 \leqslant x^{n+1} \leqslant d,\, d > 0\}$ with $\partial M \subset \partial G$. Assume that M is a hypersurface of revolution around the axis $z = x^{n+1}$ generated by the graph of a function $r(z)$, that intersects ∂G orthogonally and that encloses a region of volume v. Then the volume-preserving mean curvature flow applied to M gives a family $\{M_t\}$ of hypersurfaces of revolution around the z-axis which are generated by the graphs of functions r_t in a maximal time interval $[0,T)$. Moreover:*

1. *If $T < \infty$, then $\lim_{t \to T} \min_{x \in M_t} r_t(x) = 0$ [21, Thm. 5.26].*

2. *If $\lim_{t \to T} r_t(x_0) = 0$, then $x_0 = \lim_{t \to \infty} x_t$ with $\dot{r}_t(x_t) = 0$ [21, §5.10].*

3. *The number $N(t)$ of zeroes of $\dot{r}_t$ is finite and $N(t)$ is non-increasing.*

The volume-preserving flow can be continued in the following way: at every point where the singular graph r_T touches the z-axis, insert a cylinder of revolution C around the z-axis of small radius, so that the bases B_1, B_2 are contained in the singular region M_T enclosed by the profile curve r_T, and no other disc inside the cylinder satisfies this property. Let S be the open slab determined by the bases. Now solve the following isoperimetric problem: minimize $\mathcal{P}(\Omega, S)$ for $\Omega \subset C$ such that B_1, $B_2 \subset \overline{\Omega}$, $V(\Omega) = V(M_T \cap C)$. This problem has a solution Ω and, by Steiner symmetrization, Ω is a set of revolution around the z-axis [100]. The solution is smooth in the interior of the cylinder and merely $C^{1,1}$ near the boundary of C [115]. It can be proven that solutions are either connected, and composed of a piece of Delaunay hypersurface tangent to the boundary of the cylinder with precisely one minimum of the distance to the z-axis, or disconnected and composed of two spherical caps of the same mean curvature enclosing each one of the bases of C and, possibly, a part of the boundary of C. Replacing $M_T \cap C$ by the isoperimetric solution and performing some geometrical constructions, we can continue the flow reducing the boundary area while keeping constant the volume enclosed.

Chapter 4

Some applications to hyperbolic geometry

4.1 Introduction

Consider a compact, connected, orientable 3-manifold with a hyperbolic metric. By Mostow's Rigidity Theorem [14, Ch. C], the hyperbolic metric is unique up to isometry. Hence the geometric invariants of the metric, such as the volume or the injectivity radius, are topological invariants. The recent proof of Thurston's Geometrization Conjecture based on the results by Perelman [78], [79], [80], has increased greatly the interest in hyperbolic 3-manifolds. Let us remark that on a compact Riemann surface of genus $g \geqslant 2$, the space of hyperbolic metrics up to isotopy is the Teichmüller space, a real analytic manifold of real dimension $6g - 6$; see [14, Ch. B].

In this section we shall give two applications of the theory of isoperimetric inequalities to hyperbolic geometry. The first one is a result by Bachman, Cooper and White [10] that provides a lower bound of the genus of a Heegaard splitting of a closed orientable hyperbolic 3-manifold in terms of the injectivity radius of the manifold. The second one is a result by Adams and Morgan [2], namely the description of the isoperimetric profile of cusped $(n + 1)$-dimensional hyperbolic manifolds for small values of the volume.

Let us remark that in the above cited paper by Adams and Morgan [2], a quite complete description of the isoperimetric regions in compact hyperbolic surfaces is given. These results have been recently extended by Simonson [112].

4.2   Bounds on the Heegaard genus of a hyperbolic manifold

A *handlebody* is a 3-manifold homeomorphic to a closed regular neighborhood of a graph in $\mathbb{R}^3$. A closed surface $S \subset M$ is a *Heegaard splitting* of M if S separates M into two handlebodies. Heegaard splittings appear frequently in Riemannian geometry. For instance, any embedded minimal surface in the 3-sphere with its standard metric determines a decomposition of the sphere into two handlebodies, and so is a Heegaard splitting of $\mathbb{S}^3$; see Lawson [62]. More generally, an embedded surface with mean curvature $H \geqslant 0$ in a 3-manifold with non-negative Ricci curvature encloses a handlebody. The *genus* of a 3-manifold M is the minimum among the genera of all Heegaard splittings of M. The *injectivity radius* of a hyperbolic 3-manifold is the radius of the largest self-tangent isometrically embedded ball in M.

Let S be a connected orientable closed surface and M a closed orientable 3-manifold. Let $\Phi \colon S \times I \to M$ be a continuous map, where $I = [0, 1]$, and define $S_t = \Phi(S \times \{t\})$. Then Φ is a *sweepout* if the following conditions are satisfied:

1. S_0 and S_1 are graphs.

2. $\Phi_* \colon H_3(S \times I, \partial(S \times I)) \to H_3(M, S_0 \cup S_1)$ is an isomorphism.

The sweepout is *smooth* if Φ is C^∞.

The following has been announced by Pitts and Rubinstein [83], [84], [85]. Colding and De Lellis [32] have provided details for parts of the argument.

Theorem 4.2.1. *Let M be a compact Riemannian 3-manifold of genus g. Then, for all $\varepsilon > 0$, there is a smooth sweepout $\Phi \colon S \times I \to M$ by surfaces of genus $g = \operatorname{genus}(S)$ and a minimal surface Σ in M of genus at most g so that*

$$A(S_t) \leqslant A(\Sigma) + \varepsilon, \qquad \text{for all } t \in I. \tag{4.1}$$

Such a sweepout is called an *almost minimal* sweepout.

Observe that, by the Gauss–Bonnet Theorem, the Gauss curvature K of a minimal surface in a 3-manifold with $K_{\mathrm{sec}} \leqslant -1$ satisfies $K = K_{\mathrm{sec}} + \kappa_1 \kappa_2 \leqslant -1$. Hence we have

$$A(\Sigma) \leqslant \int_\Sigma (-K) \, dA = 2\pi \, (2g - 2)$$

and so we can replace (4.1) by

$$A(S_t) \leqslant 2\pi \, (2g - 2) + \varepsilon, \qquad \text{for all } t \in I. \tag{4.2}$$

Assuming Theorem 4.2.1, Bachman, Cooper and White [10] prove the following result.

Theorem 4.2.2 ([10, Thm. 1.1]). *Let M be a closed, orientable 3-manifold with all sectional curvatures less than or equal to -1 and Heegaard genus g. Then*

$$g \geqslant \frac{1}{2}\left(\cosh(r) + 1\right),$$

where r denotes the radius of any isometrically embedded ball in M.

In the proof of Theorem 4.2.2, the main ingredients are Theorem 4.2.1 and the following result on isoperimetric regions in hyperbolic balls [10, Thm. 3.3].

Theorem 4.2.3. *Let Ω be a compact subset of the closed ball $B = B_r$ in $\mathbb{H}^3$. Let $S = \mathrm{int}(B) \cap \partial\Omega$. Suppose that $V(\Omega) = V(B)/2$. Then the area of S is at least as large as the area of an equatorial disc. Hence*

$$A(S) \geqslant 2\pi\left(\cosh(r) - 1\right).$$

Proof. See [10, Thm. 3.3] for an alternative proof along the lines of Ros [98] and [19]. Consider an isoperimetric region $\Omega \subset B$. Apply spherical symmetrization and replace Ω by a set of revolution Ω' around a line. The perimeter is strictly decreased unless Ω were already of revolution around some line (here we use that the boundary of Ω has constant mean curvature and cannot be tangent to a sphere in a set of positive 2-dimensional Hausdorff measure unless it coincides with the sphere). Also S must touch the boundary of B (otherwise we consider the complement of Ω and move it until it touches ∂B for the first time). Observe that spherical symmetrization produces a set that has connected intersection with any sphere ∂B_s, for $s \leqslant r$. Hence $\overline{\Omega} \cap \partial B_r$ is a proper disc of ∂B_r. We conclude that S touches the axis of revolution. By the classification of constant mean curvature surfaces of revolution in $\mathbb{H}^3$, we have that S is a spherical cap or an equatorial disc. From this the result follows easily. $\square$

From the above we can prove:

Theorem 4.2.4 ([10, Thm. 4.1]). *Suppose that M is a closed, connected, orientable 3-manifold with all sectional curvatures less than or equal to -1. Suppose there is a piecewise smooth sweepout of M by surfaces which have at most area A. Let $p \in M$ and $r = \mathrm{inj}(p)$. Then*

$$2\pi\left(\cosh(r) - 1\right) \leqslant A. \tag{4.3}$$

Proof. We just consider the hyperbolic case. Let $\Phi\colon S \times I \to M$ be the given sweepout. Suppose that $\Phi\colon S \times I \to M$ is an embedding. Let $v\colon I \to \mathbb{R}$ be defined by $v(t) = V(B \cap \Phi(S \times [0, t]))$. Then v is continuous and $v(0) = 0$, $v(1) = V(M) \geqslant V(B)$. Hence there is some t_0 such that $v(t_0) = V(B)/2$. So

$$A \geqslant A(S_{t_0} \cap B) \geqslant 2\pi\left(\cosh(r) - 1\right).$$

The general case when $\Phi\colon S \times I \to M$ is not an embedding is handled in a similar way, with a slight technical complication since one has to define the volume enclosed by $\Phi(S \times \{t\})$. $\square$

Now the proof of Theorem 4.2.2 follows from Theorems 4.2.1 and 4.2.4.

4.3   The isoperimetric profile for small volumes

An interesting question in Riemannian geometry is to distinguish between Riemannian manifolds by looking at their isoperimetric profiles. Pittet [82] was able to recognize different types of homogeneous Riemannian manifolds by looking at the asymptotic isoperimetric profile for large values of the volume.

A similar idea could be applied to hyperbolic manifolds. Could we distinguish a compact $(n + 1)$-dimensional hyperbolic manifold from a cusped $(n + 1)$-dimensional hyperbolic manifold of finite volume [14] by looking at their isoperimetric profiles?

Throughout this section, M will be an $(n + 1)$-dimensional hyperbolic manifold of finite volume. We know that M is either compact, or the ends of M are *cusps*, i.e., isometric to the warped product $V \times [0, \infty)$, where (V, g) is a flat n-dimensional space form, and the Riemannian metric is given by

$$e^{-2t}g + dt^2,$$

see [14, Prop. D.3.12]. As M has finite volume, the flat manifold associated to each end is compact. We shall say that a non-compact hyperbolic manifold M is a *cusped* manifold if all its ends are cusps. A *horosurface* in a cusp is a slice $V \times \{t\}$.

In Riemannian manifolds of finite volume, existence of isoperimetric regions for any value of the volume is guaranteed by [2, §5] or by Theorem 1.2.1. Isoperimetric solutions are smooth embedded constant mean curvature hypersurfaces except on a singular set of small Hausdorff dimension.

For cusped manifolds, we have the following:

Theorem 4.3.1 ([2, Thm. 5.1]). *Let M be a cusped $(n + 1)$-dimensional hyperbolic manifold of finite volume. There exists $\varepsilon > 0$ such that the least-perimeter enclosure of a region of volume $V \leqslant \varepsilon$ is an arbitrary collection of horosurfaces around cusps, of total area $A = nV$.*

Proof. Horosurfaces around cups have mean curvature $H = 1$. By the first variation of area and volume, sliding a horosurface out of a cusp we get $dA/dV = n$. Integrating from zero volume we have $A = nV$.

If Σ is a hypersurface with constant mean curvature $H \leqslant 1$ and the injectivity radius of M at $p \in \Sigma$ is bounded below, the area of Σ is bounded below by monotonicity [68, pp. 89–90]. This bound depends on the injectivity radius of M at p. By comparison with geodesic spheres of small volume, we conclude that there is an $\varepsilon > 0$ so that an isoperimetric solution with $V \leqslant \varepsilon$ and $|H| \leqslant 1$ is contained in the cusps.

Assume that, for some $0 < V \leqslant \varepsilon$, there is an isoperimetric solution Ω which is not contained in a cusp neighborhood. Choose it to minimize $A - nV$ (this minimum has to be strictly negative). Make a small variation of the boundary of Ω supported at its regular part so that $dV/dt < 0$. Then

$$0 \leqslant \frac{dA}{dt} - n\frac{dV}{dt} = (nH - n)\frac{dV}{dt} = n(H - 1)\frac{dV}{dt}.$$

So we obtained $H \leqslant 1$ (otherwise $dA/dt - n \, dV/dt < 0$, contradicting that Ω minimizes the functional $A - nV$). So Ω must lie deep in the cusps. Moving horosurfaces until the first contact with the boundary of Ω we obtain regular points of $\partial\Omega$. Applying the maximum principle for the constant mean curvature equation contradicts the fact that $\partial\Omega$ is not a horosurface. $\qquad\square$

Recall that, on any compact Riemannian manifold, the isoperimetric profile is asymptotically that of Euclidean space [40] of the same dimension. However, for an $(n+1)$-dimensional cusped hyperbolic manifold of finite volume, the isoperimetric profile for small volumes is linear. Hence it is possible to distinguish between the two types of hyperbolic manifolds just by looking at their isoperimetric profiles for values close to 0.

An interesting recent result concerning the isoperimetric profile for small volumes in compact manifolds has been obtained by Nardulli [72], [73], [74], who has proved that isoperimetric sets for small volume are invariant by the isometries that leave invariant their center of mass. His result is stated for compact Riemannian manifolds, although it is valid whenever there is existence of *bounded* isoperimetric regions for any value of the volume, as in homogeneous manifolds. Morgan and Johnson [71] also proved that isoperimetric sets of small volumes in a compact Riemannian manifold are asymptotically round spheres.

Bibliography

[1] U. Abresch and J. Langer, *The normalized curve shortening flow and homothetic solutions*, J. Differential Geom. **23** (1986), no. 2, 175–196. MR MR845704 (88d:53001)

[2] Colin Adams and Frank Morgan, *Isoperimetric curves on hyperbolic surfaces*, Proc. Amer. Math. Soc. **127** (1999), no. 5, 1347–1356. MR MR1487351 (99h:53080)

[3] A. D. Alexandrov, *Uniqueness theorems for surfaces in the large i*, Vestnik Leningrad Univ. Math. **11** (1956), no. 19, 5–17.

[4] F. Almgren, *Optimal isoperimetric inequalities*, Indiana Univ. Math. J. **35** (1986), no. 3, 451–547. MR MR855173 (88c:49032)

[5] ———, *Spherical symmetrization*, Proceedings of the International Workshop on Integral Functionals in the Calculus of Variations (Trieste, 1985), no. 15, 1987, pp. 11–25. MR MR934771 (89h:49037)

[6] Sigurd Angenent, *Nodal properties of solutions of parabolic equations*, Rocky Mountain J. Math. **21** (1991), no. 2, 585–592, Current directions in nonlinear partial differential equations (Provo, UT, 1987). MR MR1121527

[7] Maria Athanassenas, *Volume-preserving mean curvature flow of rotationally symmetric surfaces*, Comment. Math. Helv. **72** (1997), no. 1, 52–66. MR MR1456315 (98d:58037)

[8] ———, *Behaviour of singularities of the rotationally symmetric, volume-preserving mean curvature flow*, Calc. Var. Partial Differential Equations **17** (2003), no. 1, 1–16. MR MR1979113 (2004c:35006)

[9] Thierry Aubin, *Problèmes isopérimétriques et espaces de Sobolev*, J. Differential Geom. **11** (1976), no. 4, 573–598. MR MR0448404 (56 #6711)

[10] David Bachman, Daryl Cooper, and Matthew E. White, *Large embedded balls and Heegaard genus in negative curvature*, Algebr. Geom. Topol. **4** (2004), 31–47 (electronic). MR MR2031911 (2004m:57029)

[11] Christophe Bavard and Pierre Pansu, *Sur le volume minimal de* $\mathbf{R}^2$, Ann. Sci. École Norm. Sup. (4) **19** (1986), no. 4, 479–490. MR MR875084 (88b:53048)

[12] Vincent Bayle, *A differential inequality for the isoperimetric profile*, Int. Math. Res. Not. (2004), no. 7, 311–342. MR MR2041647 (2005a:53050)

[13] Vincent Bayle and César Rosales, *Some isoperimetric comparison theorems for convex bodies in Riemannian manifolds*, Indiana Univ. Math. J. **54** (2005), no. 5, 1371–1394. MR MR2177105 (2006f:53040)

[14] Riccardo Benedetti and Carlo Petronio, *Lectures on hyperbolic geometry*, Universitext, Springer-Verlag, Berlin, 1992. MR MR1219310 (94e:57015)

[15] Itai Benjamini and Jianguo Cao, *A new isoperimetric comparison theorem for surfaces of variable curvature*, Duke Math. J. **85** (1996), no. 2, 359–396. MR MR1417620 (97m:58046)

[16] Wilhelm Blaschke, *Vorlesungen über Differentialgeometrie und geometrische Grundlagen von Einsteins Relativitätstheorie. Band I. Elementare Differentialgeometrie*, Dover Publications, New York, N. Y., 1945, 3rd ed. MR MR0015247 (7,391g)

[17] G. Bol, *Isoperimetrische Ungleichungen für Bereiche auf Flächen*, Jber. Deutsch. Math. Verein. **51** (1941), 219–257. MR MR0018858 (8,338h)

[18] E. Bombieri, E. De Giorgi, and E. Giusti, *Minimal cones and the Bernstein problem*, Invent. Math. **7** (1969), 243–268. MR MR0250205 (40 #3445)

[19] Yu. D. Burago and V. A. Zalgaller, *Geometric inequalities*, Grundlehren der Mathematischen Wissenschaften [Fundamental Principles of Mathematical Sciences], vol. 285, Springer-Verlag, Berlin, 1988, Translated from the Russian by A. B. Sosinskiĭ, Springer Series in Soviet Mathematics. MR MR936419 (89b:52020)

[20] Antonio Cañete and Manuel Ritoré, *The isoperimetric problem in complete annuli of revolution with increasing Gauss curvature*, Proc. Royal Society Edimburgh **138** (2008), no. 5, 989–1003.

[21] Esther Cabezas-Rivas, *Volume Preserving Curvature Flows in Rotationally Symmetric Spaces*, Ph.D. thesis, Universidad de Valencia, 2008.

[22] Esther Cabezas-Rivas and Vicente Miquel, *Volume-preserving mean curvature flow of revolution hypersurfaces in a rotationally symmetric space*, Math. Z. **261** (2009), 489–510.

[23] Antonio Cañete, *Stable and isoperimetric regions in rotationally symmetric tori with decreasing Gauss curvature*, Indiana Univ. Math. J. **56** (2007), no. 4, 1629–1659. MR MR2354694

[24] Isaac Chavel, *Isoperimetric inequalities*, Cambridge Tracts in Mathematics, vol. 145, Cambridge University Press, Cambridge, 2001, Differential geometric and analytic perspectives. MR MR1849187 (2002h:58040)

[25] _____ , *Riemannian geometry*, second ed., Cambridge Studies in Advanced Mathematics, vol. 98, Cambridge University Press, Cambridge, 2006, A modern introduction. MR MR2229062 (2006m:53002)

[26] Isaac Chavel and Edgar A. Feldman, *Isoperimetric inequalities on curved surfaces*, Adv. in Math. **37** (1980), no. 2, 83–98. MR MR591721 (82i:53056)

[27] Bang-yen Chen, *On the total curvature of immersed manifolds. I. An inequality of Fenchel-Borsuk-Willmore*, Amer. J. Math. **93** (1971), 148–162. MR MR0278240 (43 #3971)

[28] Jaigyoung Choe, Mohammad Ghomi, and Manuel Ritoré, *Total positive curvature of hypersurfaces with convex boundary*, J. Differential Geom. **72** (2006), no. 1, 129–147. MR MR2215458 (2007a:53076)

[29] _____ , *The relative isoperimetric inequality outside convex domains in* $\mathbf{R}^n$, Calc. Var. Partial Differential Equations **29** (2007), no. 4, 421–429. MR MR2329803

[30] Jaigyoung Choe and Manuel Ritoré, *The relative isoperimetric inequality in Cartan-Hadamard 3-manifolds*, J. Reine Angew. Math. **605** (2007), 179–191. MR MR2338131

[31] Kai-Seng Chou and Xi-Ping Zhu, *The curve shortening problem*, Chapman & Hall/CRC, Boca Raton, FL, 2001. MR MR1888641 (2003e:53088)

[32] Tobias H. Colding and Camillo De Lellis, *The min-max construction of minimal surfaces*, arXiv:math.AP/0303305 v2, 2003.

[33] Christopher B. Croke, *A sharp four-dimensional isoperimetric inequality*, Comment. Math. Helv. **59** (1984), no. 2, 187–192. MR MR749103 (85f:53060)

[34] Lawrence C. Evans and Ronald F. Gariepy, *Measure theory and fine properties of functions*, Studies in Advanced Mathematics, CRC Press, Boca Raton, FL, 1992. MR MR1158660 (93f:28001)

[35] F. Fiala, *Le problème des isopérimètres sur les surfaces ouvertes à courbure positive*, Comment. Math. Helv. **13** (1941), 293–346. MR MR0006422 (3,301b)

[36] Bruno Franchi, Raul Serapioni, and Francesco Serra Cassano, *Rectifiability and perimeter in the Heisenberg group*, Math. Ann. **321** (2001), no. 3, 479–531. MR MR1871966 (2003g:49062)

[37] M. Gage and R. S. Hamilton, *The heat equation shrinking convex plane curves*, J. Differential Geom. **23** (1986), no. 1, 69–96. MR MR840401 (87m:53003)

[38] M. E. Gage, *Curve shortening makes convex curves circular*, Invent. Math. **76** (1984), no. 2, 357–364. MR MR742856 (85i:52004)

[39] Michael E. Gage, *Curve shortening on surfaces*, Ann. Sci. École Norm. Sup. (4) **23** (1990), no. 2, 229–256. MR MR1046497 (91a:53072)

[40] Sylvestre Gallot, *Inégalités isopérimétriques et analytiques sur les variétés riemanniales*, Société Mathématique de France, Astérisque **163–164** (1988), 31–91.

[41] Nicola Garofalo and Duy-Minh Nhieu, *Isoperimetric and Sobolev inequalities for Carnot-Carathéodory spaces and the existence of minimal surfaces*, Comm. Pure Appl. Math. **49** (1996), no. 10, 1081–1144. MR MR1404326 (97i:58032)

[42] Enrico Giusti, *Minimal surfaces and functions of bounded variation*, Monographs in Mathematics, vol. 80, Birkhäuser Verlag, Basel, 1984. MR MR775682 (87a:58041)

[43] E. Gonzalez, U. Massari, and I. Tamanini, *On the regularity of boundaries of sets minimizing perimeter with a volume constraint*, Indiana Univ. Math. J. **32** (1983), no. 1, 25–37. MR MR684753 (84d:49043)

[44] Matthew A. Grayson, *The heat equation shrinks embedded plane curves to round points*, J. Differential Geom. **26** (1987), no. 2, 285–314. MR MR906392 (89b:53005)

[45] _____, *Shortening embedded curves*, Ann. of Math. (2) **129** (1989), no. 1, 71–111. MR MR979601 (90a:53050)

[46] Misha Gromov, *Metric structures for Riemannian and non-Riemannian spaces*, Progress in Mathematics, vol. 152, Birkhäuser Boston Inc., Boston, MA, 1999, Based on the 1981 French original [MR0682063 (85e:53051)], With appendices by M. Katz, P. Pansu and S. Semmes, Translated from the French by Sean Michael Bates. MR MR1699320 (2000d:53065)

[47] G. H. Hardy, J. E. Littlewood, and G. Pólya, *Inequalities*, Cambridge, at the University Press, 1952, 2nd ed. MR MR0046395 (13,727e)

[48] Philip Hartman, *Geodesic parallel coordinates in the large*, Amer. J. Math. **86** (1964), 705–727. MR MR0173222 (30 #3435)

[49] Laurent Hauswirth, Joaquín Pérez, Pascal Romon, and Antonio Ros, *The periodic isoperimetric problem*, Trans. Amer. Math. Soc. **356** (2004), no. 5, 2025–2047 (electronic). MR MR2031051 (2004j:53014)

[50] Ernst Heintze and Hermann Karcher, *A general comparison theorem with applications to volume estimates for submanifolds*, Ann. Sci. École Norm. Sup. (4) **11** (1978), no. 4, 451–470. MR MR533065 (80i:53026)

[51] F. Hélein, *Isoperimetric inequalities and calibrations*, Progress in partial differential equations: the Metz surveys, 4, Pitman Res. Notes Math. Ser., vol. 345, Longman, Harlow, 1996, pp. 92–105. MR MR1394707 (97d:53062)

[52] Frédéric Hélein, *Inégalité isopérimétrique et calibration*, Ann. Inst. Fourier (Grenoble) **44** (1994), no. 4, 1211–1218. MR MR1306553 (95i:53068)

[53] Wu-teh Hsiang and Wu-Yi Hsiang, *On the uniqueness of isoperimetric solutions and imbedded soap bubbles in noncompact symmetric spaces. I*, Invent. Math. **98** (1989), no. 1, 39–58. MR MR1010154 (90h:53078)

[54] Wu-Yi Hsiang, *Isoperimetric regions and soap bubbles*, Differential geometry, Pitman Monogr. Surveys Pure Appl. Math., vol. 52, Longman Sci. Tech., Harlow, 1991, pp. 229–240. MR MR1173044 (93h:49060)

[55] _____, *On soap bubbles and isoperimetric regions in noncompact symmetric spaces. I*, Tohoku Math. J. (2) **44** (1992), no. 2, 151–175. MR MR1161609 (93a:53044)

[56] Bruce Kleiner, *An isoperimetric comparison theorem*, Invent. Math. **108** (1992), no. 1, 37–47. MR MR1156385 (92m:53056)

[57] Gary Lawlor, *Metacalibrations*, work in progress.

[58] _____, *A new minimization proof for the brachistochrone*, Amer. Math. Monthly **103** (1996), no. 3, 242–249. MR MR1376179

[59] _____, *Proving area minimization by directed slicing*, Indiana Univ. Math. J. **47** (1998), no. 4, 1547–1592. MR MR1687102 (2001b:49057)

[60] Gary Lawlor and Frank Morgan, *Paired calibrations applied to soap films, immiscible fluids, and surfaces or networks minimizing other norms*, Pacific J. Math. **166** (1994), no. 1, 55–83. MR MR1306034 (95i:58051)

[61] _____, *Curvy slicing proves that triple junctions locally minimize area*, J. Differential Geom. **44** (1996), no. 3, 514–528. MR MR1431003 (98a:53012)

[62] H. Blaine Lawson, Jr., *The unknottedness of minimal embeddings*, Invent. Math. **11** (1970), 183–187. MR MR0287447 (44 #4651)

[63] Peter Li and Shing Tung Yau, *A new conformal invariant and its applications to the Willmore conjecture and the first eigenvalue of compact surfaces*, Invent. Math. **69** (1982), no. 2, 269–291. MR MR674407 (84f:53049)

[64] Sebastián Montiel, *Unicity of constant mean curvature hypersurfaces in some Riemannian manifolds*, Indiana Univ. Math. J. **48** (1999), no. 2, 711–748. MR 2001f:53131

[65] Sebastián Montiel and Antonio Ros, *Compact hypersurfaces: the Alexandrov theorem for higher order mean curvatures*, Differential geometry, Pitman Monogr. Surveys Pure Appl. Math., vol. 52, Longman Sci. Tech., Harlow, 1991, pp. 279–296. MR MR1173047 (93h:53062)

[66] Frank Morgan, *Clusters minimizing area plus length of singular curves*, Math. Ann. **299** (1994), no. 4, 697–714. MR MR1286892 (95g:49083)

[67] _____, *Regularity of isoperimetric hypersurfaces in Riemannian manifolds*, Trans. Amer. Math. Soc. **355** (2003), no. 12, 5041–5052 (electronic). MR MR1997594 (2004j:49066)

[68] _____, *Geometric Measure Theory*, 4th ed., Academic Press Inc., San Diego, CA, 2008, A beginner's guide.

[69] _____, *The Levy-Gromov isoperimetric inequality in convex manifolds with boundary*, J. Geom. Anal. **18** (2008), no. 4, 1053–1057. MR MR2438911

[70] Frank Morgan, Michael Hutchings, and Hugh Howards, *The isoperimetric problem on surfaces of revolution of decreasing Gauss curvature*, Trans. Amer. Math. Soc. **352** (2000), no. 11, 4889–4909. MR MR1661278 (2001b:58024)

[71] Frank Morgan and David L. Johnson, *Some sharp isoperimetric theorems for Riemannian manifolds*, Indiana Univ. Math. J. **49** (2000), no. 3, 1017–1041. MR MR1803220 (2002e:53043)

[72] Stefano Nardulli, *Le profil isopérimétrique d'une variété riemannienne compact pour les petits volumes*, Ph.D. thesis, Université d'Orsay, 2006.

[73] Stefano Nardulli, *The isoperimetric profile of a compact riemannian manifold for small volumes*, Ann. Glob. Anal. Geom. **36** (2009), 111–131.

[74] _____, *Regularity of solutions of the isoperimetric problem that are close to a smooth manifold*, arXiv:0710.1849, 2007.

[75] Robert Osserman, *The isoperimetric inequality*, Bull. Amer. Math. Soc. **84** (1978), no. 6, 1182–1238. MR MR0500557 (58 #18161)

[76] Pierre Pansu, *Sur la régularité du profil isopérimétrique des surfaces riemanniennes compactes*, Ann. Inst. Fourier (Grenoble) **48** (1998), no. 1, 247–264. MR MR1614957 (99i:53035)

[77] Renato H. L. Pedrosa and Manuel Ritoré, *Isoperimetric domains in the Riemannian product of a circle with a simply connected space form and applications to free boundary problems*, Indiana Univ. Math. J. **48** (1999), no. 4, 1357–1394. MR MR1757077 (2001k:53120)

[78] Grisha Perelman, *The entropy formula for the Ricci flow and its geometric applications*, arXiv:math.DG/0211159 v1, 2002.

[79] _____, *Finite extinction time for the solutions to the Ricci flow on certain three-manifolds*, arXiv:math.DG/0307245 v1, 2003.

[80] _____, *Ricci flow with surgery on three-manifolds*, arXiv:math.DG/0303109 v1, 2003.

[81] Peter Petersen, *Riemannian geometry*, 2nd ed., Graduate Texts in Mathematics, vol. 171, Springer, New York, 2006. MR MR2243772 (2007a:53001)

[82] Ch. Pittet, *The isoperimetric profile of homogeneous Riemannian manifolds*, J. Differential Geom. **54** (2000), no. 2, 255–302. MR MR1818180 (2002g:53088)

[83] Jon T. Pitts and J. H. Rubinstein, *Existence of minimal surfaces of bounded topological type in three-manifolds*, Miniconference on geometry and partial differential equations (Canberra, 1985), Proc. Centre Math. Anal. Austral. Nat. Univ., vol. 10, Austral. Nat. Univ., Canberra, 1986, pp. 163–176. MR MR857665 (87j:49074)

[84] ———, *Applications of minimax to minimal surfaces and the topology of 3-manifolds*, Miniconference on geometry and partial differential equations, 2 (Canberra, 1986), Proc. Centre Math. Anal. Austral. Nat. Univ., vol. 12, Austral. Nat. Univ., Canberra, 1987, pp. 137–170. MR MR924434 (89a:57001)

[85] ———, *Equivariant minimax and minimal surfaces in geometric three-manifolds*, Bull. Amer. Math. Soc. (N.S.) **19** (1988), no. 1, 303–309. MR MR940493 (90a:53014)

[86] Murray H. Protter and Hans F. Weinberger, *Maximum principles in differential equations*, Prentice-Hall Inc., Englewood Cliffs, N.J., 1967. MR MR0219861 (36 #2935)

[87] Manuel Ritoré, *Applications of compactness results for harmonic maps to stable constant mean curvature surfaces*, Math. Z. **226** (1997), no. 3, 465–481. MR MR1483543 (98m:53082)

[88] ———, *Examples of constant mean curvature surfaces obtained from harmonic maps to the two sphere*, Math. Z. **226** (1997), no. 1, 127–146. MR MR1472144 (98h:53098)

[89] ———, *Index one minimal surfaces in flat three space forms*, Indiana Univ. Math. J. **46** (1997), no. 4, 1137–1153. MR MR1631568 (99h:58041)

[90] ———, *Constant geodesic curvature curves and isoperimetric domains in rotationally symmetric surfaces*, Comm. Anal. Geom. **9** (2001), no. 5, 1093–1138. MR MR1883725 (2003a:53018)

[91] ———, *The isoperimetric problem in complete surfaces of nonnegative curvature*, J. Geom. Anal. **11** (2001), no. 3, 509–517. MR MR1857855 (2002f:53109)

[92] ———, *Optimal isoperimetric inequalities for three-dimensional Cartan-Hadamard manifolds*, Global theory of minimal surfaces, Clay Math. Proc., vol. 2, Amer. Math. Soc., Providence, RI, 2005, pp. 395–404. MR MR2167269 (2006f:53043)

[93] _____, *A proof by calibration of an isoperimetric inequality in the Heisenberg group*, arXiv:0803.1313, 2008.

[94] Manuel Ritoré and Antonio Ros, *Stable constant mean curvature tori and the isoperimetric problem in three space forms*, Comment. Math. Helv. **67** (1992), no. 2, 293–305. MR MR1161286 (93a:53055)

[95] _____, *The spaces of index one minimal surfaces and stable constant mean curvature surfaces embedded in flat three manifolds*, Trans. Amer. Math. Soc. **348** (1996), no. 1, 391–410. MR MR1322955 (96f:58038)

[96] Manuel Ritoré and César Rosales, *Existence and characterization of regions minimizing perimeter under a volume constraint inside Euclidean cones*, Trans. Amer. Math. Soc. **356** (2004), no. 11, 4601–4622 (electronic). MR MR2067135 (2005g:49076)

[97] Antonio Ros, *Compact hypersurfaces with constant higher order mean curvatures*, Rev. Mat. Iberoamericana **3** (1987), no. 3-4, 447–453. MR 90c:53160

[98] _____, *The isoperimetric problem*, Global theory of minimal surfaces, Clay Math. Proc., vol. 2, Amer. Math. Soc., Providence, RI, 2005, pp. 175–209. MR MR2167260 (2006e:53023)

[99] _____, *Stable periodic constant mean curvature surfaces and mesoscopic phase separation*, Interfaces Free Bound. **9** (2007), no. 3, 355–365. MR MR2341847 (2008h:58025)

[100] César Rosales, *Isoperimetric regions in rotationally symmetric convex bodies*, Indiana Univ. Math. J. **52** (2003), no. 5, 1201–1214. MR MR2010323 (2004h:58018)

[101] Erhard Schmidt, *Über das isoperimetrische Problem im Raum von n Dimensionen*, Math. Z. **44** (1939), 689–788.

[102] _____, *Die isoperimetrischen Ungleichungen auf der gewöhnlichen Kugel und für Rotationskörper im n-dimensionalen sphärischen Raum*, Math. Z. **46** (1940), 743–794. MR MR0003733 (2,262e)

[103] _____, *Über die isoperimetrische Aufgabe im n-dimensionalen Raum konstanter negativer Krümmung. I. Die isoperimetrischen Ungleichungen in der hyperbolischen Ebene und für Rotationskörper im n-dimensionalen hyperbolischen Raum*, Math. Z. **46** (1940), 204–230. MR MR0002196 (2,12e)

[104] _____, *Über eine neue Methode zur Behandlung einer Klasse isoperimetrischer Aufgaben im Grossen*, Math. Z. **47** (1942), 489–642. MR MR0016219 (7,527h)

[105] _____, *Beweis der isoperimetrischen Eigenschaft der Kugel im hyperbolischen und sphärischen Raum jeder Dimensionenzahl*, Math. Z. **49** (1943), 1–109. MR MR0009127 (5,106d)

[106] Felix Schulze, *Nonlinear evolution by mean curvature and isoperimetric inequalities*, J. Differential Geom. **79** (2008), no. 2, 197–241.

[107] H. A. Schwarz, *Beweis des Satzes, dass die Kugel kleinere Oberfläche besitzt, als jeder andere Körper gleichen Volumens*, Nachrichten von der Königlichen Gesellschaft der Wissenchaften und der Georg-Augusts-Universität zu Göttingen (1884), no. 1, 1–13.

[108] Katsuhiro Shiohama, Takashi Shioya, and Minoru Tanaka, *The geometry of total curvature on complete open surfaces*, Cambridge Tracts in Mathematics, vol. 159, Cambridge University Press, Cambridge, 2003. MR MR2028047 (2005c:53037)

[109] Katsuhiro Shiohama and Minoru Tanaka, *The length function of geodesic parallel circles*, Progress in differential geometry, Adv. Stud. Pure Math., vol. 22, Math. Soc. Japan, Tokyo, 1993, pp. 299–308. MR MR1274955 (95b:53054)

[110] Leon Simon, *Lectures on geometric measure theory*, Proceedings of the Centre for Mathematical Analysis, Australian National University, vol. 3, Australian National University Centre for Mathematical Analysis, Canberra, 1983. MR MR756417 (87a:49001)

[111] _____, *Existence of surfaces minimizing the Willmore functional*, Comm. Anal. Geom. **1** (1993), no. 2, 281–326. MR MR1243525 (94k:58028)

[112] Matthew Simonson, *The isoperimetric Problem on Euclidean, Spherical, and Hyperbolic Surfaces*, Master senior thesis, Williams College, 2008.

[113] Carlo Sinestrari, *Formation of singularities in the mean curvature flow*, CRM Advanced Course on Geometric Flows and Hyperbolic Geometry, 2008.

[114] Edward Stredulinsky and William P. Ziemer, *Area minimizing sets subject to a volume constraint in a convex set*, J. Geom. Anal. **7** (1997), no. 4, 653–677. MR MR1669207 (99k:49089)

[115] _____, *Area minimizing sets subject to a volume constraint in a convex set*, J. Geom. Anal. **7** (1997), no. 4, 653–677. MR MR1669207 (99k:49089)

[116] Bernd Süssmann, *Curve shortening flow and the Banchoff-Pohl inequality on surfaces of nonpositive curvature*, Beiträge Algebra Geom. **40** (1999), no. 1, 203–215. MR MR1678540 (2000b:53098)

[117] Peter Topping, *Mean curvature flow and geometric inequalities*, J. Reine Angew. Math. **503** (1998), 47–61. MR MR1650335 (99m:53080)

[118] _____, *The isoperimetric inequality on a surface*, Manuscripta Math. **100** (1999), no. 1, 23–33. MR MR1714389 (2000i:53107)

Advanced Courses in Mathematics CRM Barcelona

Edited by
Carles Casacuberta

Since 1995 the Centre de Recerca Matemàtica (CRM) has organised a number of Advanced Courses at the post-doctoral or advanced graduate level on forefront research topics in Barcelona. The books in this series contain revised and expanded versions of the material presented by the authors in their lectures.

Argyros, S. / Todorcevic, S.
Ramsey Methods in Analysis (2005)
ISBN 978-3-7643-7264-4

Audin, M. / Cannas da Silva, A. /
Lerman, E.
Symplectic Geometry of Integrable Hamiltonian Systems (2003)
ISBN 978-3-7643-2167-3

Bertoluzza, S. / Falletta, S. / Russo, G. / Shu, C.-W.
Numerical Solutions of Partial Differential Equations (2008). ISBN 978-3-7643-8939-0

The first part of this volume is devoted to the use of wavelets to derive some new approaches in the numerical solution of PDEs, showing in particular how the possibility of writing equivalent norms for the scale of Besov spaces allows to develop some new methods. The second part provides an overview of the modern finite-volume and finite-difference shock-capturing schemes for systems of conservation and balance laws, with emphasis on providing a unified view of such schemes by identifying the essential aspects of their construction. In the last part a general introduction is given to the discontinuous Galerkin methods for solving some classes of PDEs, discussing cell entropy inequalities, nonlinear stability and error estimates.

Brady, N. / Riley, T. / Short, H.
The Geometry of the Word Problem for Finitely Generated Groups (2006)
ISBN 978-3-7643-7949-0

Brown, K.A. / Goodearl, K.R.
Lectures on Algebraic Quantum Groups (2002)
ISBN 978-3-7643-6714-5

Catalano, D. / Cramer, R. / Damgård, I. /
Di Creszenso, G. / Pointcheval, D. / Takagi, T.
Contemporary Cryptology (2005)
ISBN 978-3-7643-7294-1

Christopher, C. / Li, C.
Limit Cycles of Differential Equations (2007)
ISBN 978-3-7643-8409-8

Cohen, R.L. / Hess, K. / Voronov, A.A.
String Topology and Cyclic Homology (2006)
ISBN 978-3-7643-2182-6

Da Prato, G.
Kolmogorov Equations for Stochastic PDEs (2004)
ISBN 978-3-7643-7216-3

Drensky, V. / Formanek, E.
Polynomial Identity Rings (2004)
ISBN 978-3-7643-7126-5

Dwyer, W.G. / Henn, H.-W.
Homotopy Theoretic Methods in Group Cohomology (2001). ISBN 978-3-7643-6605-6

Geroldinger, A. / Ruzsa, I.Z.
Combinatorial Number Theory and Additive Group Theory (2009). ISBN 978-3-7643-8961-1

Additive combinatorics is a relatively recent term coined to comprehend the developments of the more classical additive number theory, mainly focussed on problems related to the addition of integers. Some classical problems like the Waring problem on the sum of k-th powers or the Goldbach conjecture are genuine examples of the original questions addressed in the area. One of the features of contemporary additive combinatorics is the interplay of a great variety of mathematical techniques, including combinatorics, harmonic analysis, convex geometry, graph theory, probability theory, algebraic geometry or ergodic theory. This book gathers the contributions of many of the leading researchers in the area and is divided into three parts. The two first parts correspond to the material of the main courses delivered, Additive combinatorics and non-unique factorizations, by A. Geroldinger, and Sumsets and structure, by I.Z. Ruzsa. The third part collects the notes of most of the seminars which accompanied the main courses, and which cover a reasonably large part of the methods, techniques and problems of contemporary additive combinatorics.

Markvorsen, S. / Min-Oo, M.
Global Riemannian Geometry: Curvature and Topology (2003)
ISBN 978-3-7643-2170-3

Mislin, G. / Valette, A.
Proper Group Actions and the Baum-Connes Conjecture (2003)
ISBN 978-3-7643-0408-9

Myasnikov, A. / Shpilrain, V. / Ushakov, A.
Group-based Cryptography (2008)
ISBN 978-3-7643-8826-3

Ritoré, M. / Sinestrari, C.
Mean Curvature Flow and Isoperimetric Inequalities (2009)
ISBN 978-3-0346-0212-9

BIRKHÄUSER

Printed in the United States
By Bookmasters